Essentials of Inorganic Chemistry 1

D. M. P. Mingos

Sir Edward Frankland BP Professor of Inorganic Chemistry,
Imperial College

OXFORD   NEW YORK   TORONTO
OXFORD UNIVERSITY PRESS
1995

Oxford University Press, Walton Street, Oxford OX2 6DP

Oxford   New York
Athens   Auckland   Bangkok   Bombay
Calcutta   Cape Town   Dar es Salaam   Delhi
Florence   Hong Kong   Istanbul   Karachi
Kuala Lumpur   Madras   Madrid   Melbourne
Mexico City   Nairobi   Paris   Singapore
Taipei   Tokyo   Toronto

and associated companies in
Berlin   Ibadan

Oxford is a trade mark of Oxford University Press

Published in the United States
by Oxford University Press Inc., New York

A catalogue record for this book is available from the British Library

Library of Congress Cataloging in Publication Data
Mingos, D. M. P.
Essentials of inorganic chemistry/D. M. P. Mingos.
(Oxford chemistry primers; 28–)
Includes bibliographical references
1. Chemistry, Inorganic.   I. Title.   II. Series: Oxford chemistry primers; 28, etc.

QD151.2.M56   1995   546'.03—dc20   94-46277
ISBN   0 19 855848 1

Typeset by the author and AMA Graphics Ltd., Preston, Lancs
Printed in Great Britain on acid-free paper by
The Bath Press, Bath, Avon

Series Editor's Foreword

Oxford Chemistry Primers are designed to give a concise introduction to all chemistry students by providing the material that would usually form an 8–10 lecture course. As well as providing up-to-date information, this series provides explanations and rationales that form the framework of an understanding of inorganic chemistry.

Mike Mingos has presented an alternative style of book giving a glossary of key terms and concepts within the subject. The choice of items is relevant to anyone starting an undergraduate course in chemistry and this book will be an essential companion to other text books. The rapid access to the core concepts will be invaluable to all students. This is achieved with the style one would expect from a writer who himself has contributed substantially to the way we think about inorganic chemistry.

John Evans
Department of Chemistry, University of Southampton

Preface

A university lecturer always faces the problem of choosing a level which is appropriate for that specific course. This problem is particularly acute for first year undergraduates because of variations in the contents of pre-university courses. The aim of this book is to assist the transition between school and university by providing a concise account of the concepts in inorganic chemistry which either should have been covered in a pre-university course or will be introduced in the first year. The topics have been arranged alphabetically for easy access although the specific entries are hopefully sufficiently detailed to give more than a dictionary definition. The page limitations imposed by the Oxford Chemistry Primer series have required me to choose those topics which I thought essential for freshman undergraduates and I hope that my choice is representative. There are plans to adopt the same format for a companion volume which will deal with more advanced topics suitable for second and third year students—*Non-essential concepts in inorganic chemistry* perhaps?

I should like to thank the many people who have helped me complete this book either by contributing to the production or reading parts of the manuscript. In particular Simon Cotton and John Evans who read the draft version of the manuscript and made many helpful suggestions, David Goodgame who read the proofs and Jeff Leigh who carefully read the section dealing with nomenclature. My son Adam also provided a consumers' view of the material. On the production side Neil Hyatt started the process of producing the camera-ready copy, but the tremendous burden of coming to grips with the formatting of the document and ensuring that the book did not exceed 92 pages fell on Jack Barrett. I cannot thank him enough for the dedication and patience he showed completing the task. The responsibility for the inaccuracies that remain in the book is regretfully mine alone.

London
April 1994

D.M.P.M.

Contents

Acids and bases — Brønsted definition

In the Brønsted definition acids (AH) and bases (B) function as proton donors and acceptors in a complementary manner defined by the equation:

$$AH \;+\; B \;\rightleftharpoons\; A^- \;+\; BH^+$$

On the right hand side of the equation where proton transfer has been achieved the roles of A and B are reversed. In the reverse reaction the BH^+ ion is a Brønsted acid, the A^- ion being a Brønsted base. The BH^+ ion is the **conjugate acid** of the base B and the A^- ion is the **conjugate base** of the acid AH.

Brønsted acid		Brønsted base		Conjugate base		Conjugate acid
H_3O^+	$+$	NH_3	$\rightleftharpoons$	H_2O	$+$	NH_4^+
$[Fe(H_2O)_6]^{3+}$	$+$	H_2O	$\rightleftharpoons$	$[Fe(H_2O)_5OH]^{2+}$	$+$	H_3O^+
H_2SO_4	$+$	H_2O	$\rightleftharpoons$	HSO_4^-	$+$	H_3O^+

Compounds which are able to function both as Brønsted acids and bases are described as amphoteric. Water is a particularly important example.

AH		B		A^-		BH^+
HNO_3	$+$	H_2O	$\rightleftharpoons$	NO_3^-	$+$	H_3O^+
H_2O	$+$	$HN{=}C(NH_2)_2$	$\rightleftharpoons$	OH^-	$+$	$H_2N{=}C(NH_2)_2^+$

Neutralization reactions occur when the conjugate acid and conjugate base of a molecule combine.

$$NH_4^+ \;+\; NH_2^- \;\rightleftharpoons\; 2NH_3$$

acid   base   neutralization product

$$H_3SO_4^+ \;+\; HSO_4^- \;\rightleftharpoons\; 2H_2SO_4$$

The reverse of such reactions represents the autoionization processes commonly observed in protic solvents.

$$2H_2O \;\rightleftharpoons\; H_3O^+ \;+\; OH^-$$

$$2H_2SO_4 \;\rightleftharpoons\; H_3SO_4^+ \;+\; HSO_4^-$$

$$2NH_3 \;\rightleftharpoons\; NH_4^+ \;+\; NH_2^-$$

These reactions have equilibria which lie well to the left-hand side. The equilibrium constants for these processes are described as autoprotolysis constants, e.g. that for water (also known as the ionic product) is

$$K_{AP} \;=\; [H_3O^+][OH^-] \;=\; 1.0 \times 10^{-14} \text{ at } 25°C$$

	pK_{AH}
HClO	7.2
HClO$_2$	2.0
HClO$_3$	−1.0
HClO$_4$	−10.0

Acids become progressively stronger as the number of Cl=O groups increases. These allow resonance stabilization of the negative charge formed on oxygen when the oxo-anion is formed from the acid (*see Resonance*).

	pK_{AH}
H$_3$PO$_2$	2
H$_3$PO$_3$	1.8
H$_3$PO$_4$	2.1

Number of P=O groups stays constant, therefore acid strengths similar.

	pK_B
NH$_2$OH	7.97
NH$_2$NH$_2$	5.77
NH$_3$	4.74
MeNH$_2$	3.36
Me$_2$NH	3.29
Me$_3$N	4.28

The strengths of Brønsted bases are represented by K_B the values of which are related to the acidity constant K_{BH+} of their conjugate acids by the relationship:

$$pK_B = 14 - pK_{BH+}$$

Electron withdrawing groups decrease basicity and electron donating groups increase the basicity relative to NH$_3$.

In pure water $[H_3O^+] = [OH^-] = (1.0 \times 10^{-14})^{1/2} = 1.0 \times 10^{-7}$ at 25°C and pH = $-\log_{10}[H_3O^+]$ = 7. It follows that in acidic solutions $[H_3O^+] > [OH^-]$, pH < 7, and in alkaline solutions $[OH^-] > [H_3O^+]$, pH > 7.

Autoprotolysis constants for other inorganic solvents are:

H$_2$SO$_4$ pK_{AP} = 2.9 (25°C)

NH$_3$ pK_{AP} = 27.7 (−50°C)

The relative strengths of Brønsted acids are defined by the acidity constant, K_{AH}:

$$K_{AH} = [H_3O^+][A^-]/[AH]$$

Since acidity constants vary over a wide range of values it is convenient to express them logarithmically as pK values:

$$pK_{AH} = -\log_{10}K_{AH}$$

The greater the degree of dissociation of the acid (at a given concentration) the larger is its acidity constant, K_{AH}, and the smaller is its pK_{AH} value. Some typical values for common inorganic acids are:

	pK_{AH}		pK_{AH}		pK_{AH}
HI	~ −9	H$_2$SO$_4$	~ −3	H$_2$SO$_3$	1.9
HBr	~ −8	HNO$_3$	~ −1.4	H$_2$SeO$_3$	2.6
HCl	~ −7	H$_3$PO$_4$	2.1	H$_2$TeO$_3$	2.7
HF	~ 3.2	HNO$_2$	3.3	H$_2$CO$_3$	3.9

The complementary definitions of Brønsted acids and their conjugate bases mean that the pK_{AH} values also provide a scale for defining the base strengths of Brønsted bases. In particular a strong Brønsted acid AH has a weak conjugate base and its pK_{AH} value is small or even negative. A strong base A$^-$ has a weak conjugate acid AH with a pK_{AH} value which is large and positive.

Acids and bases — Lewis definition

G.N. Lewis (1923) defined acids and bases in terms of electron-pair donation and acceptance. A *base* is an electron-pair donor and *acid* is an electron-pair acceptor. Examples:

Me$_3$N: + BF$_3$ →   Me$_3$N: → BF$_3$

Lewis base   Lewis acid   Adduct or complex

2NH$_3$ + Ag$^+$ → [Ag(NH$_3$)$_2$]$^+$ (H$_3$N:→ Ag$^+$←:NH$_3$)

The new shared bond formed as a result of donation of an electron pair from the donor is described as a *co-ordinate bond* or *dative bond* and the resulting molecule may be described as an *adduct* or a *complex*. The donor of the electron pair may be described as a *ligand* and may be anionic, e.g. Cl$^-$ or neutral, e.g. NH$_3$.

The empty acceptor orbital may be either a p orbital as in BF$_3$, or empty s orbitals (e.g. H$^+$), d orbitals, e.g. Co^{2+} (Co^{II}), or f orbitals, e.g. U^{6+} (U^{VI}).

The smaller the cation and the larger the charge/size ratio the stronger is the metal ion as a Lewis acid towards Lewis bases such as ammonia and water.

If a molecule has both a donor orbital and an empty orbital it may function as a Lewis base and Lewis acid, e.g. SO_2.

Lewis basicity trends

Group trends

Down a group of the Periodic Table Lewis basicity generally parallels basicity towards proton provided only σ-bonds are formed and steric effects are not important.

	Molar heat of reaction with BMe_3	Product with HCl
Me_3N	-74 kJ mol^{-1}	sublimes at $250\,°C$
Me_3P	-67 kJ mol^{-1}	sublimes at $125\,°C$
Me_3As	exists only at $-80°C$	unstable at room temperature
Me_3Sb	no compound formed	unstable at $-80\,°C$

The donor atom follows the electronegativity trend: $Me_3N > Me_2O > MeF$. Acceptor towards O and N donors $B > Be > Li$; $Be > Mg > Ca$.

Effect of substituents

Base strength: $Me_3N > H_3N > F_3N$.
Acid strength: $F_3B > H_3B > Me_3B$.
However, $Br_3B > Cl_3B > F_3B$ and $Me_3B > (MeO)_3B$
because of π–bonding effects (see *Dative bond*).

Steric effects

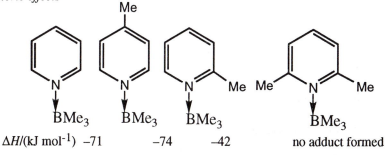

$\Delta H/$(kJ mol^{-1}) -71 -74 -42 no adduct formed

ΔH refers to the enthalpy change for $Me_3B + L \rightarrow Me_3B{:}L$

Hard and soft acids and bases

Co-ordination chemists established from equilibrium constant data that the later transition metals and the heavier post-transition metals (Cu^+, Ag^+, Au^+, Cd^{2+}, Hg^{2+}, Pd^{2+}, Pt^{2+}, known as Class B metal ions) formed more stable complexes with the donor atoms P, S, and I, i.e.

$$N \ll P > As > Sb;\ O \ll S > Se \sim Te;\ F < Cl < Br < I$$

Hard acids

H^+, Li^+, Na^+, K^+

Be^{2+}, Mg^{2+}, Ca^{2+}, Sr^{2+}

Al^{3+}, La^{3+}, Ce^{4+}, U^{4+}, UO_2^{2+}

Ti^{4+}, VO^{2+}, Cr^{3+}, Mn^{2+}, Fe^{3+}

BF_3, $AlCl_3$, Ga^{3+}, In^{3+}

Borderline acids

Fe^{2+}, Co^{2+}, Ni^{2+}, Cu^{2+}, Zn^{2+}

Rh^{3+}, Ir^{3+}, Ru^{3+}, Os^{2+}

NO^+, SO_2

Soft acids

Pd^{2+}, Pt^{2+}, Pt^{4+}

Cu^+, Ag^+, Au^+

Cd^{2+}, Hg^{2+}, $HgCH_3^+$

Hard bases

NH_3, H_2O, OH^-, O^{2-}

NO_3^-, CO_3^{2-}, SO_4^{2-}

Borderline bases

N_3^-, N_2, SO_3^{2-}

Soft bases

H^-, R^-, C_2H_4, CN^-, RNC

CO, SCN^-, R_3P, R_2S, RS^-

$S_2O_3^{2-}$, I^-

The actinides

Element	Z	Element	Z
Th	90	Bk	97
Pa	91	Cf	98
U	92	Es	99
Np	93	Fm	100
Pu	94	Mv	101
Am	95	No	102
Cm	96	Lr	103

Elements of groups 1 and 2		
n	ns^1	ns^2
2	Li	Be
3	Na	Mg
4	K	Ca
5	Rb	Sr
6	Cs	Ba
7	Fr	Ra

In contrast the earlier transition metals and the lighter post-transition metals (Al^{3+}, Co^{3+}, Fe^{3+}, Ti^{4+}, known as Class A metals) formed more stable complexes according to the following orders of donor atoms:

$$N \gg P > As > Sb; F > Cl > Br > I; O > S > Se > Te$$

Pearson suggested that these empirical facts could be generalized into the 'hard' and 'soft' acid and base principle whereby, **hard bases** such as nitrogen, oxygen and fluorine ligands bind more strongly to **hard acids** such as Al^{3+}, Co^{3+}, Fe^{3+}, etc. and **soft bases** such as phosphorus, sulfur, and iodine ligands bind more strongly to **soft acids** such as Cu^+, Ag^+, Au^+, Cd^{2+}, Hg^{2+}.

In this context hard and soft are taken to be qualitative measures of the polarizabilities of the acids and bases.

The usefulness of the hard and soft acid and base generalization arises from its ability to predict the correct sign for the enthalpy change, ΔH, of equilibria represented by:

$$h_a s_b + s_a h_b \rightleftharpoons h_a h_b + s_a s_b \qquad \Delta H < 0$$

where h and s are hard and soft acids and bases respectively. For example,

$$CaS + CdO \rightleftharpoons CaO + CdS \qquad \Delta H = -66 \text{ kJ mol}^{-1}$$

$$BaI_2 + HgF_2 \rightleftharpoons BaF_2 + HgI_2 \qquad \Delta H = -397 \text{ kJ mol}^{-1}$$

or the equilibrium constant, K_{eq}, for

$$HOHgCH_3 + SO_3H^- \rightleftharpoons HOH + CH_3HgSO_3^- \qquad K > 1$$

Actinides

Elements with atomic number 90 (Thorium) to 103 (Lawrencium). (See *Periodic table.*) All the elements are radioactive, those with $Z \geq 93$ (Np) have been produced artificially by nuclear reactions.

Alkali metals and alkaline earth metals

The alkali metals are the elements of group 1 of the periodic table (Li, Na, K, Rb, Cs, and Fr). The alkaline earth metals are the elements of group 2 of the periodic table (Be, Mg, Ca, Sr, Ba, and Ra). See *Periodic table.*

Allotropy

Many elements exist in more than one form in the same state. At its simplest these forms represent different crystalline modifications of the same basic molecular structure, e.g. monoclinic and rhombic sulphur both have S_8 rings, but they are packed differently in the monoclinic and rhombic crystalline forms. In more complex examples the different allotropes have markedly different structures, e.g. diamond, graphite, and the recently discovered cage molecules including C_{60} and C_{70}.

If the two allotropes are in equilibrium then this is described as dynamic allotropy.

Other examples of allotropy:

Oxygen O_2 (dioxygen) O_3 (ozone, trioxygen)

In some cases allotropes exist only over specific temperature ranges. For example tin undergoes the following transition:

grey α-tin	13.2°C	white β-tin	161°C	γ-tin
diamond structure	→	metallic	→	metallic
				distorted close packed

(This type of allotropy is also called enantiotropy.)

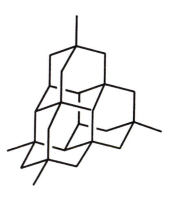

Diamond

Alternation effect

For the majority of the elements in the periodic table the properties of compounds with the same formula change regularly down a column. However, for the post-transition elements an alternation in properties is observed. SeF_6 is less stable than suggested by a simple interpolation between SF_6 and TeF_6. This is a thermodynamic effect and most clearly seen in the following enthalpies of formation:

	SF_6	SeF_6	TeF_6
$\Delta H^{\ominus}{}_f/kJ\ mol^{-1}$	−1209	−1029	−1318

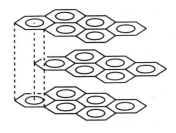

Graphite

It can also be seen in a more qualitative fashion when known compounds of these elements are compared:

$NF_5{}^*$	PF_5	AsF_5	SbF_5	BiF_5
$NCl_5{}^*$	PCl_5	$AsCl_5{}^{**}$	$SbCl_5$	$BiCl_5{}^*$
$NBr_5{}^*$	PBr_5	$AsBr_5{}^*$	$SbBr_5$	$BiBr_5{}^*$

* not known   ** stable only below −50°C

Or related compounds which have very different stabilities:

SO_3 SeO_3 (unstable) TeO_3

Buckminsterfullerene

The thermodynamic origins of the effect may be traced to the ionization potentials of the atoms which do not decrease in a regular fashion down the column but alternate because of the intervention of the d and f shells (see *Ionization energy*).

Amorphous

A solid which is not sufficiently crystalline to show an X-ray powder diffraction pattern or appear as crystals under an electron microscope is described as amorphous. The crystal structures of amorphous solids therefore do not contain unit cells ordered in the same precise manner throughout the solid, i.e. they do not have long-range order.

Anhydrous

A salt or compound which does not contain water and in particular does not contain water in its crystal lattice.

Aprotic solvent

A solvent which is neither a proton donor nor acceptor under normal conditions. For example, CH_3CN is generally considered an aprotic solvent, but is a protophile in the presence of concentrated H_2SO_4 and protogenic in the presence of a strong base such as KO^tBu.

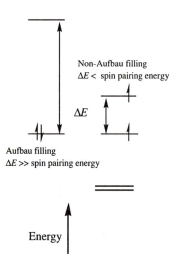

Non-Aufbau filling
$\Delta E <$ spin pairing energy

ΔE

Aufbau filling
$\Delta E \gg$ spin pairing energy

Energy

Three parallel spin electrons in a three-fold degenerate level.

Aufbau principle

The energy states of an atom or a molecule are quantized and defined by quantum numbers. The *Aufbau* (German: 'building up') principle states that the lowest energy (most stable) orbitals are occupied first and in a manner consistent with the Pauli exclusion principle, i.e. with two electrons with opposite spins in each singly degenerate orbital. When orbitals of the same energy (degenerate orbitals) are available the electrons occupy the empty orbitals singly, before pairing (see *Exchange energy*). If an orbital is *n*-fold degenerate then it can accommodate a total of $2n$ electrons. The application of the Aufbau principle to atoms leading to the structure of the periodic table is discussed under *Periodic table* and the electronic configurations which result are summarized on the inside back cover.

Exceptions to the Aufbau principle may occur when orbitals have similar energies and electron repulsion effects favour the presence of a single electron in each orbital rather than pairing in the lowest energy orbital (see *screening* and *penetration* under *Periodic table*).

The Aufbau principle is also applicable to the filling of *molecular orbitals* which allows the correct prediction of the ground state electron configurations of molecules.

Back donation

See *Dative bond*.

B

Base metal

A common and relatively inexpensive metal which corrodes or tarnishes in air, e.g. Fe, Pb, or Cu.

Bond energies

The *Bond dissociation energy* (D_0) for a diatomic covalent molecule, A_2, is the change in internal energy, ΔU_0, for the process

$$A_2(g) \longrightarrow 2A(g)$$

at 0 K, where the parent molecule and product atoms are in their ground electronic states. A more useful quantity is the *bond dissociation enthalpy* at the standard temperature of 25°C, $\Delta H^{\ominus}$. A table of typical values is given in the margin.

For polyatomic molecules there is more than one bond dissociation energy and one must be careful to specify exactly the dissociation process being referred to. Generally values of the *average bond enthalpy, E* (also known as the *bond energy term*), which are based on the enthalpies for dissociation processes at 298 K, are quoted.

$$AX_n(g) \longrightarrow A(g) + nX(g) \quad \Delta U \quad E = \Delta U/n$$

Examples of calculations:

$$SiF_4(g) \longrightarrow Si(g) + 4F(g) \qquad \Delta H_{298} = 2384 \text{ kJ mol}^{-1}$$
$$E(Si\text{-}F) = 2384/4 = 596 \text{ kJ mol}^{-1}$$

$$Mo(CO)_6(g) \longrightarrow Mo(g) + 6CO(g) \qquad \Delta H_{298} = 908 \text{ kJ mol}^{-1}$$
$$E(Mo\text{-}CO) = 908/6 = 151 \text{ kJ mol}^{-1}$$

Although average bond enthalpies are not exactly transferable from one molecule to another most values are sufficiently similar to be able to make meaningful chemical predictions.

For example, $E(P\text{-}Cl)$ in $PCl_5 = 259$ kJ mol^{-1}
$E(P\text{-}Cl)$ in $PCl_3 = 322$ kJ mol^{-1}

If it is assumed that $E(P\text{-}Cl) \sim 300$ kJ mol^{-1} some semi-quantitative conclusions can be made concerning the relative stabilities of phosphorus compounds containing P–Cl bonds. In molecules with more than one sort of bond, e.g. H_2O_2, the average bond enthalpies may only be estimated by making some assumptions about the constancy of bond strengths, for example, that $E(OH)$ in H_2O_2 is the same as $E(OH)$ in H_2O. Some typical values of $E(X\text{-}Y)$ are given (in kJ mol^{-1}) in the table in the margin.

Molecule	$\Delta H^{\ominus}$ dissoc kJ mol^{-1}
H_2	436
F_2	158
O_2	496
N_2	946
CO	1075

Typical average bond enthalpies (E/kJ mol^{-1})	
C-H; 412	C-C; 348
Si-H; 318	C=C; 612
N-H; 388	C≡C; 837
P-H; 322	N-N; 163
O-H; 463	N=N; 409
S-H; 338	N≡N; 946
F-H; 562	P-P; 172
Cl-H; 431	S-S; 264
Br-H; 366	Cl-Cl; 242
Si-Si; 226	Si-O; 374

Some general trends:
1. X–H bond enthalpies usually decrease down the Group for most of the post transition metals, e.g. C–H > Si–H > Ge–H > Sn–H

2. Similar trends for other bonds providing that neither atom has lone pairs then the maximum tends to occur for the second long period, P–F > As–F > N–F and P–O > As–O > N–O

3. For halogens and noble gases the opposite trend is observed, I–F > Br–F > Cl–F

4. Element–element bond enthalpies generally decrease down the column for post transition elements and increase for a column of transition elements.

(continued on page 8)

5. Element–element multiple bonds are strongest for the second period elements, O=O > S=S and N≡N > P≡P.

6. For transition elements the average bond enthalpies for M–L bonds generally increase down a group.

Bond order

Bond order is the prediction from theoretical considerations of the relative strength of a bond. Classically the bond order is the number of electron pairs being shared between connected atoms. Integer increases in bond order are observed in many series of related molecules, e.g. as in the carbon–carbon bonded compounds shown in the Table below. In a molecular orbital analysis the formal bond order is equal to (no. of electrons in bonding molecular orbitals – no. of electrons in antibonding molecular orbitals)/2. If a molecular orbital is non-bonding then the formal bond order does not change as this orbital is populated by electrons. It is generally observed that a bond becomes stronger and shorter as the bond order increases.

	$H_3C–CH_3$	$H_2C=CH_2$	$HC≡CH$
Bond order	1	2	3
Bond length (pm)	154	135	120
Bond enthalpy (kJ mol^{-1})	348	612	837
C–C force constant (N m^{-1})	450	843	1902

Molecule	No. of valence electrons
B_2	6
C_2	8
N_2	10
O_2	12
F_2	14
Ne_2	16

The relationships between formal bond order, bond enthalpy, bond length and force constant for the diatomic molecules, B_2–N_2, are illustrated below.

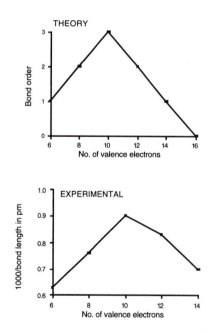

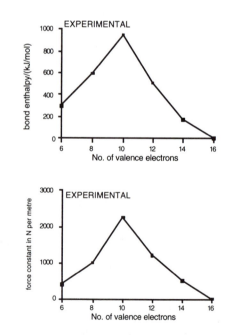

Fractional bond orders may arise either because the molecule is a radical with an odd number of electrons or because of resonance delocalization effects. Examples of diatomics with fractional bond orders are shown on the next page.

	N_2	O_2^+	O_2	O_2^-	O_2^{2-}
Bond order	3	2.5	2	1.5	1
Bond length(pm)	110	112	121	130	148
Dissociation energy (kJ mol^{-1})	941	643	494	395	155
Force constant (N m^{-1})	2290	1632	1180	620	383

The bonding of the nitrate(III) ion, NO_2^-, can be formulated as one N=O double bond with a single bond between the nitrogen atom and an oxygen atom carrying a negative charge, N–O$^-$: the two N–O bonds are identical with a length of 124 pm, compared with the observed lengths of 147 pm for the single N–O bond in NH_2OH and 115 pm for the bond in NO which has an order of 2.5. The double bond character (see *Resonance*) and the negative charge are shared by all three atoms in the ion. The order of both N–O bonds is 1.5.

Born–Haber cycle

A cycle of reactions based on the first law of thermodynamics or *Hess's law* which is used to calculate the lattice energy of an ionic compound. An overall reaction (e.g. the formation of an ionic compound) may be divided into stages, for each of which the enthalpy change is either known or may be calculated. The main uses of these cycles are in the calculation of the lattice energies of compounds or, when the lattice energies may be calculated theoretically, in the estimation of the enthalpies of formation of compounds. (see also *Lattice energy, Ionic bond, Ionization energy,* and *Electron affinity.*)

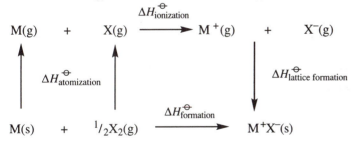

The first law of thermodynamics (or Hess's law) determines that:

$$\Delta H^{\ominus}_{\text{formation}} = \Delta H^{\ominus}_{\text{atomization}} + \Delta H^{\ominus}_{\text{ionization}} + \Delta H^{\ominus}_{\text{lattice formation}}$$

$$\Delta H^{\ominus}_{\text{atomization}} = \Delta H^{\ominus}_{\text{sublimation}} (M,s) + \tfrac{1}{2} \Delta H^{\ominus}_{\text{dissociation}} (X_2,g)$$

$$\Delta H^{\ominus}_{\text{ionization}} = \text{Ionization energy } (M,g) - \text{Electron affinity } (X,g)$$

$$\text{Electron affinity } (X,g) = -\Delta H^{\ominus}_{\text{electron capture}} (X,g)$$

C

Canonical form

A theoretical resonance form of a molecule used in *valence bond theory* (see *Resonance*).

Catalyst

A substance which increases the rate of a chemical reaction without being consumed in the reaction. At the molecular level the catalyst lowers the activation energy by participating in a cyclical series of reactions each of which has a low activation energy. The catalyst is regenerated at the end of each cycle of reactions.

An autocatalytic chemical reaction is one in which one of the products functions as a catalyst. Such a reaction has a rate which increases with time since the catalyst concentration becomes progressively larger.

Catenation

Catenation is the ability of an element to form a wide range of molecular compounds containing element–element bonds. Carbon provides the prime example where chain, ring, and polyhedral molecules occur in great variety based on single, double, and triple carbon–carbon bonds. In general those elements which form strong bonds in the element itself (see Table) also form a wide range of catenated compounds, e.g. P, Si, transition elements of groups 6, 7, and 8.

Bond	E (kJ mol^{-1})
C–C	345
Si–Si	222
Ge–Ge	188
Sn–Sn	146
B–B	293
Al–Al	186
Ga–Ga	113
In–In	100
N–N	167
P–P	201
As–As	146
Sb–Sb	121

Chalcogens

The chalcogens are the group 16 elements (O, S, Se, Te, Po). (See *Periodic table*.)

Group16	elements
n	ns^2np^4
2	O
3	S
4	Se
5	Te
6	Po

Chelate

See *Ligand*.

Clathrate, inclusion compound, microporous solid

A complex in which one component (the host) forms a crystal lattice containing open tunnels or channels which can accommodate molecules of the second component (the guest molecules). No chemical bonds are formed between host and guest and only van der Waals forces operate.

Closest packing of spheres

Molecules and atoms combine to form crystalline solids because of the attractive forces that exist between them. For metal atoms the forces result from the overlap of partially filled orbitals and therefore the attractive forces are generally strong and the metals have correspondingly high melting points. For inert gas atoms the van der Waals forces are weak and their melting points are very low. For spherical atoms such as these the forces are not directional and the most stable structure results if the atoms adopt the most densely packed arrangement which minimizes the distances between atoms. The most economical way of packing spheres of equal size is called *closest packing*.

In two dimensions spheres may be closest packed by having each sphere touching six neighbouring spheres at the points of a regular hexagon. Six voids are generated between the central sphere and the adjacent spheres, but only three of these may be occupied when a second layer is built on top of the first. The voids may therefore be labelled XYXYXY around the central sphere. If a two-layer structure is built by putting all the spheres in all the X sites, then an interesting choice results when one comes to build a third layer above the second layer. If the X sites are occupied then this layer lies exactly above the first layer and the structure is described as *hexagonal close packed* (hcp). If the Y sites are occupied then this layer is out of register with respect to the second layer by 60° and the structure is described as *cubic close packed* (ccp). Both sphere packings are equally efficient at occupying sphere space and in total 74.05% ($\pi/(3\sqrt{2})$) of the space is occupied by the spheres.

The Table below provides examples of close-packed structures of the metallic elements. The group 18 gases also form ccp solids. Molecules which are approximately spherical also adopt close packed structures, e.g. SF_6. Of the 94 elements whose structures are known in the solid state 53 adopt closest sphere packings under normal conditions.

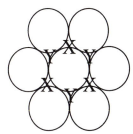

Below, three atoms of a second layer added to first layer in Y positions. The shaded atom belongs to the third layer and is located dirctly above an atom positioned in the first layer. The XYXY positions of successive layers is known as hexagonal closest packing (hcp). The layers are described as ABAB...

Li	Be												
i	h												
Na	Mg											Al	
i	h											c	
K	Ca	Sc	Ti	V	Cr	Mn	Fe	Co	Ni	Cu	Zn	Ga	
i	c	h	h	i	i	#	i	h	c	c	h*	#	
Rb	Sr	Y	Zr	Nb	Mo	Tc	Ru	Rh	Pd	Ag	Cd	In	Sn
i	c	h	h	i	i	h	h	c	c	c	h*	c*	#
Cs	Ba	La	Hf	Ta	W	Re	Os	Ir	Pt	Au	Hg	Tl	Pb
i	i	hc	h	i	i	h	h	c	c	c	c*	h	c
Fr	Ra	Ac											
	i	c											

Ce	Pr	Nd	Pm	Sm	Eu	Gd	Tb	Dy	Ho	Er	Tm	Yb	Lu
c	hc	hc	hc	hhc	i	h	h	h	h	h	h	c	h
Th	Pa	U	Np	Pu	Am	Cm	Bk	Cf	Es	Fm	Md	No	Lr
c	#	#	#	#	hc	hc	c,h, c						

h = hexagonal close packing; c = cubic close packing;

hc, hhc = other stacking varieties of close packing;

i = body centred cubic packing; # = structure type of its own;

* = distorted.

In the diagram below the first two layers are arranged AB as in hcp but the third layer (indicated by the shaded atom) is placed so that the X voids (still visible in hcp) are blocked out. The third layer is denoted by C and the structure based upon an ABCABC arrangement is known as cubic closest packed (ccp).

In both ccp and hcp the sphere packings result in co-ordination numbers of 12. In ccp the co-ordination polyhedron is the cuboctahedron and in hcp the anticuboctahedron (see *Co-ordination geometry*).

The idealized packings described above assume that once the packing is set by the first three layers then the same pattern is maintained for all

subsequent layers. This is generally the case, although there are some examples amongst the lanthanides where the layer packings involve stacking sequences derived from both hcp and ccp.

If one starts in two dimensions with a square rather than a hexagonal array then the type of packing which results is described as *body centred cubic*. The packing efficiency is reduced to 68.02% ($\pi\sqrt{3}/8$). Around each atom there are 8 nearest neighbours arranged at the vertices of a cube with a further 6 neighbours only 15% more distant originating from the atoms at the centres of adjacent cubes. Therefore, the co-ordination number is ambiguous and good arguments could be made for either 8 or 14 co-ordination.

The body centred cubic packing is less common than the close packed structures amongst the elements, but 15 elements nevertheless adopt this structure.

The structures of metals do not follow simple periodic trends, but the following generalizations are helpful. The ccp structure is favoured for metals with filled or nearly filled electron shells, e.g. the Cu, Ni, and Co family, Ca and Sr; bcc is the primary structure for half-filled shells, s^1 group 1 metals, s^2d^5 and s^1d^5 for transition elements, f^7 for Eu. The hcp structure is formed by metals with intermediate electron configurations.

The assumption that atoms have spherically symmetrical force fields breaks down in some cases causing distortions from ideal structures. The following elements: Ga, Sn, Bi, B, Mn, Pa, U, Np, and Pu have distorted structures. Indeed Ga has a structure closely related to that found for elemental iodine with each gallium atom having one close neighbour and six which are more distant.

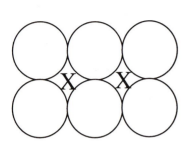

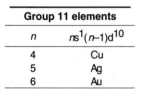

The second layer atoms are positioned in the A voids of the first layer:

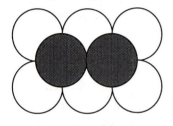

Group 11 elements	
n	$ns^1(n-1)d^{10}$
4	Cu
5	Ag
6	Au

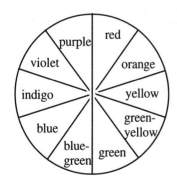

Coinage metals

The coinage metals are the group 11 elements of the periodic table, Cu, Ag, and Au (see *Periodic table*).

Colour

A compound is coloured if it absorbs in a region of the spectrum of the incident white light. Visible light consists of the perceived colour ranges: red (720–640 nm), orange (640–590 nm), yellow (590–530 nm), green (530–490 nm), blue (490–420 nm), and violet (420–400 nm). If the compound has a band in one of these regions the compound will show the complementary colour. For example if a compound absorbs strongly in the red region of the spectrum it will appear blue or blue-green. If a compound has strong absorptions throughout the visible region it is black. The 'artist's wheel' shown may be used to predict the position of the absorption band which is diametrically opposite to the observed colour.

Origin of colour

Absorption of radiation in the wavelength range 700–400nm (170–300 kJ mol^{-1}) corresponding to the visible region causes transitions of electrons between energy levels within the molecule.

The precise form of the spectrum in the visible region depends on the number and position of the electronic transitions, their intensities, and the breadth of the bands. The number of bands may be frequently predicted from quantum mechanics and depends on the symmetry of the molecule and the number of energy levels which are accessible when light is absorbed by the molecule. The intensity of a band is a molecular quantity – the molar extinction coefficient (or molar absorption coefficient), whose magnitude is defined by whether the transition is allowed or forbidden. Electronic transitions within a particular manifold of orbitals, e.g. d $\rightarrow$ d and f $\rightarrow$ f transitions in transition metal and lanthanide complexes respectively are generally weak because they are forbidden.

According to the Laporte rule an allowed transition must involve a transition from orbitals with different symmetry properties with respect to inversion, i.e. gerade (g) to ungerade (u) transitions or vice versa.

Since d orbitals are gerade then a d $\rightarrow$ d transition is g to g and forbidden. Similarly f $\rightarrow$ f transitions are forbidden. Transitions which are allowed arise from molecular orbitals with gerade symmetry to molecular orbitals with ungerade symmetry or *vice versa*. If the original molecular orbital and the final molecular orbital are equally delocalized over the molecule the transition is described as a molecular transition, e.g. the colour of the halogens Cl_2, Br_2, and I_2 result from such molecular transitions. If the original and final molecular orbitals are localized in different parts of the molecule then the transition is described as a charge transfer transition or a mixed valence transition. The purple colour of MnO_4^- results from a charge transfer transition since the electron is localized in a molecular orbital which is concentrated on the oxygen atoms and after the absorption of light the electron is localized mainly on the manganese atom. The blue colour of Prussian blue results from a transition of an electron from an iron(II) site to an iron(III) site in $Fe_4[Fe(CN)_6]_3.14H_2O$.

The breadth of a band depends on the extent of mixing between the electronic transition and vibrational modes of a molecule. In general if the ground and excited electronic states of a molecule have very different bonding character then the band is broad because the electronic transition is associated with a wide range of vibronic states. In contrast if the ground and excited states have similar bonding character then the band is narrow. These narrow bands are very important in the design of lasers and are the basis, for example, of the ruby laser.

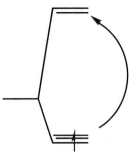

promotion of electrons between energy levels

The inversion operation: the orbital must be identical at the points (x,y,z) and (–x,–y,–z) to be gerade (g) otherwise it is ungerade (u).

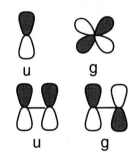

Laporte selection rule
g $\rightarrow$ u and u $\rightarrow$ g allowed
g $\rightarrow$ g and u $\rightarrow$ u forbidden.

Spin selection rule
$\Delta S = 0$, i.e. the spin of the electron must not change during the transition.

Complex

See *Acids and bases* and *Co-ordination compound*.

Configuration

An electron configuration is the arrangement of electrons occupying the energy levels in an atom either in its ground or excited state (see inside back cover).

A molecular configuration is the arrangement of groups or atoms in a molecule.

Conformation

The specific shape of a molecule resulting from the rotation of one part of a molecule about a single bond.

Congener

Elements in the same group of the periodic table, e.g. phosphorus is a congener of nitrogen in group 15 of the periodic table.

Co-ordination compound

A positively charged central ion (or possibly a neutral atom) surrounded in a symmetrical manner by a shell of ions or molecules described as *ligands*.

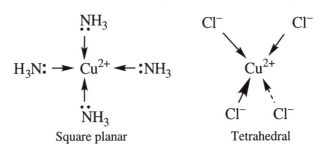

Square planar   Tetrahedral

Co-ordination geometry

The idealized polyhedral arrangements of the ligand atoms around the central metal atom in a co-ordination compound are used to describe its geometry. The common co-ordination numbers are in the range 2–9 and the most commonly observed co-ordination polyhedral (or for flat molecules polygonal) geometries are illustrated together with relevant examples.

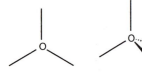

1. Trigonal planar   2. Tetrahedral

3. Square planar   4. Trigonal bipyramidal

Co-ordination polyhedra	Example in molecular chemistry	Example in solid state infinite structures
Trigonal planar (1)	BF_3 and $[Cu(CN)_3]^{2-}$	O in TiO_2
Tetrahedral (2)	CCl_4 and $Ni(CO)_4$	Zn and S in ZnS
Square planar(3)	XeF_4 and $[PdCl_4]^{2-}$	Pt in PtS and Cu and O in Nd_2CuO_4
Trigonal bipyramidal (4)	PF_5 and $[CdCl_5]^{3-}$	Very rare

Numbers refer to the illustrated co-ordination polyhedra

By changing either the number of valence electrons or ligands alternative geometries may be observed. When the energies of alternative geometries are similar a whole continuum of structures may be observed, e.g. 5-co-ordination. When the alternative geometries have very similar energies they may be interconverted by vibrational modes and they are described as fluxional or stereochemically non-rigid.

Co-ordination polyhedra	Example in molecular chemistry	Example in solid state infinite structures
Square pyramidal (5)	$WOCl_4$ and $[Ni(CN)_5]^{3-}$	V_2O_5
Octahedral (6)	SF_6 and $[TiF_6]^{2-}$	Na and Cl in NaCl
Trigonal prism (7)	$W(CH_3)_6$ and $[Re(S_2C_2Ph_2)_3]$	Mo in MoS_2 and As in NiAs
Pentagonal bipyramid (8)	IF_7 and $[V(CN)_7]^{4-}$	Very rare
Capped octahedron (9)	$[Mo(CNPh)_7]^{2+}$	Very rare
Capped trigonal prism (10)	$[NbF_7]^{2-}$	Very rare
Cube	$[PaF_8]^{3-}$	Cs in CsCl
Dodecahedron	$[Mo(CN)_8]^{4-}$	U in UCl_4
Square antiprism	$[TaF_8]^{3-}$	U in UF_4
Tricapped trigonal prism	$[ReH_9]^{2-}$	U in UCl_3
Cuboctahedron	Very rare	Cubic close packed metals eg: Ni
Anticuboctahedron	Very rare	Hexagonal close packed metals eg: Ti

Numbers refer to the illustrated co-ordination polyhedra

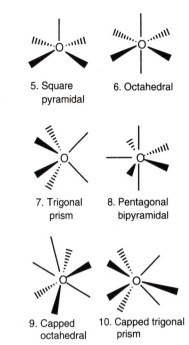

5. Square pyramidal

6. Octahedral

7. Trigonal prism

8. Pentagonal bipyramidal

9. Capped octahedral

10. Capped trigonal prism

Co-ordination number

The number of ligands which are co-ordinated to a central metal ion. For transition metals common co-ordination numbers are 2, 4, and 6, although examples of 3, 5, 7, 8, and 9 co-ordination are also well established. The lanthanides generally have higher co-ordination numbers than the transition metals, e.g. $[Mn(H_2O)_6]^{2+}$ but $[Ce(H_2O)_9]^{3+}$.

The concept may become ambiguous when some of the bonds in the molecule are longer than others: e.g. $[Cu(NH_3)_6]^{2+}$, a distorted octahedron, has four Cu–N bonds at 207 pm and two Cu–N bonds at 262 pm.

The concept is also used to describe the number of atoms surrounding an ion in the solid state structure of an ionic compound: e.g. solid NaCl has six Cl^- ions surrounding each Na^+ and six Na^+ ions surrounding each Cl^-.

Core and valence electrons

Core electrons are those electrons which are never utilized in chemical bonding. Their high ionzation energies and contracted nature mean that they are not perturbed by the orbitals of neighbouring atoms. In general core electrons may be defined as all those electrons in orbitals which are associated with the noble gas prior to that element in the periodic table.

Example: the electronic configuration of a cobalt atom ($Z = 27$) in the ground state is $1s^2 2s^2 2p^6 3s^2 3p^6 4s^2 3d^7$. The core electrons are: $1s^2 2s^2 2p^6 3s^2 3p^6$ since this is the electronic configuration of the previous noble gas element argon, in its ground state. The valence electrons are $4s^2 3d^7$. This allows Co to be represented as $[Ar]4s^2 3d^7$ (see inside back cover).

Cuboctahedron

Anti-cuboctahedron

Valence electrons

The valence electrons as defined above do not necessarily all become involved in bonding in all compounds of that element. For example cobalt does not form any compounds in the +9 oxidation state corresponding to the utilization of all electrons, (cf. Mn(+7) in MnO_4^-); it does, however, form compounds in oxidation states (−1) to (+5).

Valence orbitals

The valence orbitals of an atom are defined as those which are occupied by the valence electrons and those which are empty but are accessible by electron promotion processes involving energies comparable with those involved in chemical bond formation.

	Valence orbitals	Maximum no. of two-centre two-electron bonds
Groups 1 and 2, 12–17	ns, np	4
Groups 3–11	ns, np and nd	9
Group 18 (noble gases)	ns (for He), ns, np for Ne to Rn	
Lanthanides	6s,4f,5d (6p)	
Actinides	7s,5f,6d	

The valence orbitals set an upper limit on the number of two-centre two-electron bonds which can be formed by that element. Additional bonds may be formed using three-centre four-electron bonds. For the second long row of the periodic table the 3d orbitals lie above the 3s and 3p orbitals and therefore potentially could be used for bond formation. However, the consensus of opinion is that the promotion energy required to promote electrons from 3s or 3p into 3d is too large for these orbitals to make a significant contribution to bonding. Therefore, molecules such as PF_5, SF_6, XeF_2, XeF_4, and XeF_6 are generally discussed in terms of three-centre four-electron bonding schemes, which ignore the contribution of the d orbitals.

The valence orbitals also define the type of *hybrid orbitals* which may be utilized by an element in its compounds. For example:

For a more detailed discussion see *Hypervalent*.

Groups 1 & 2, 12 −17	sp, sp^2 and sp^3 hybridization schemes
Transition elements	sp^3 in d^{10} complexes such as $Ni(CO)_4$
	sp^2d in d^8 complexes such as $PtCl_4^{2-}$
	sp^3d in d^8 complexes such as $Fe(CO)_5$
	sp^3d^2 in d^6 complexes such as $Mo(CO)_6$
	sp^3d^3 in d^4 complexes such as $[NbF_7]^{2-}$
	sp^3d^4 in d^2 complexes such as $[Mo(CN)_8]^{4-}$
	and sp^3d^5 in d^0 complexes such as $[ReH_9]^{2-}$

Covalent bond

A pair of electrons shared by two neighbouring atoms is described as a *covalent bond* and represented by a line, e.g. O–H, Si–H etc. Since the electron density is concentrated in the region between the nuclei this type of bond is described as a sigma (σ) bond.

Atomic orbitals

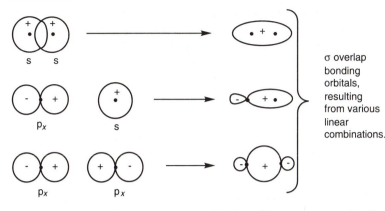

σ overlap bonding orbitals, resulting from various linear combinations.

More than one pair of electrons may be shared between atoms leading to a *multiple bond.* In the following series the nitrogen–nitrogen bond order is progressively increased from 1 to 3:

e.g. $H_2N–NH_2$ F–N=N–F   N≡N

    single σ   double σ + π   triple σ + 2π

The π-bonds have a nodal plane, i.e. a region of zero electron density in the plane containing the atoms.

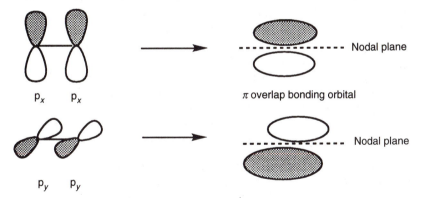

π overlap bonding orbital

In transition metal chemistry the availability of d orbitals permits the formation of quadruple bonds, σ + 2π + δ, where the δ bond has two nodal planes in the plane containing the atoms forming the bond.

Diagrams for the σ overlap of two d_{z^2} orbitals, the π overlap of two d_{xz} (or two d_{yz}) orbitals, and the δ overlap of two d_{xy} orbitals are shown on page 18. The bonding and structure of the $[Re_2Cl_8]^{2-}$ ion is shown in the margin as an example of δ bonding. The order of the rhenium–rhenium bond is four. The σ bond is formed by the overlap of the two d_{z^2} orbitals. The two π bonds are formed by the overlap of the two pairs of d_{xz} and d_{yz} orbitals. The δ bond is formed by the overlap of the two d_{xy} orbitals. The remaining $d_{x^2-y^2}$ orbitals are used in the Re–Cl bonding. The quadruple bond ensures that the two sets of four chloride ligands have an eclipsed rather than a staggered conformation.

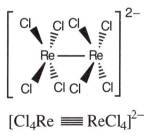

$$[Cl_4Re \equiv ReCl_4]^{2-}$$

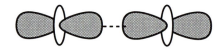

d$_\sigma$ overlap

no nodal planes along bond

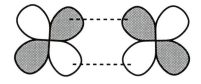

d$_\pi$ overlap

1 nodal plane along bond

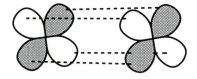

d$_\delta$ overlap

2 nodal planes along bond

The number of covalent bonds which an atom is capable of forming is described as its *valency.* The majority of atoms exhibit multiple valencies corresponding to alternative ways of utilizing their valence electrons.

Example: S ($3s^23p^4$) exhibits valencies of 2 in SH_2, 4 in SF_4, and 6 in SF_6.

Since atoms do not have equal *electronegativities* and the sharing of electrons pairs may not occur exclusively with one atom the classical covalent bond description provided above is over-restrictive. In *valence bond theory* these deficiencies are overcome by the concept of *resonance.*

Examples: The polar bonds in HF may be represented by the following resonance forms:

H—F ↔ H$^+$ F$^-$

The multiple bond in NO_2^- by those shown in the margin.

In molecular orbital theory, the wave functions representing the electron pair are *delocalized* over the molecule and therefore these problems are not encountered. Nevertheless, it is sometimes convenient to re-express the molecular orbitals in localized bond terms in order to make a connection with classical bonding theory.

The strength of a covalent bond depends upon the the extent of overlap between orbitals, electronegativity difference, co-ordination number, inter-atomic lone pair–lone pair repulsions, hybridization, and steric effects (see *Bond energies* and *Bond order*).

Crystal field theory

See *Ligand field theory*.

Cryptand

See *Ligand*.

Crystal structures

The main crystal structure types may be summarized as follows:

Description	Structural description	Physical characteristics	Examples
Molecular	Molecules interacting through van der Waals and hydrogen bonding forces. usually each molecule is surrounded by approx. 12 identical molecules (see *Inter-molecular forces*).	Soft, low m.p. crystals. Insulators. Large coefficient of expansion.	All molecular inorganic and organic compounds, C_6H_6, $B_3N_3H_6$, SF_6, S_8.
Extended covalent	Covalently bonded three dimensional infinite polymers (see below).	Hard, high m.p. crystals. Insulators.	Diamond, SiO_2.
Metallic	Metal–metal bonds between atoms which form a close packed arrangement usually (see *Closest packing of spheres*).	Soft, m.p. varies from below room temp. (Hg) to over $3000°C$, good conductors, strength depends on structural defects and strengths of metal–metal bonds.	Fe, Co, Al, Na.
Ionic	Cations and anions held together in an infinite three dimensional array by electrostatic forces .	Hard, brittle, high m.p., insulators. Conductors when molten.	NaCl, CsCl.

Infinite structures

Some elements and compounds exist in the crystalline state with infinite structures. Their complete description requires a definition of the local co-ordination geometries and the relative disposition of local co-ordination polyhedra within the infinite three-dimensional structure. The important structural classes adopted by elements are given below (see *Closest packing of spheres*).

Structure type	Co-ordination polyhedron of each atom
Cubic close packed	12-cuboctahedron (12-co)
Hexagonal close packed	12-anti-cuboctahedron (12-aco)
Body centred cubic	8-cubic (8-c)
Simple cubic	6-octahedral (6-o)
Diamond structure	4-tetrahedral (4-t)

The primitive cubic unit cell

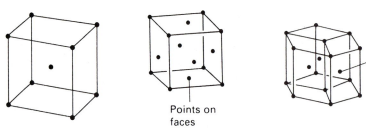

body centred cubic   cubic close packed (fcc)   hexagonal close packed

Homeotypic structures are based upon these prototypes but the co-ordination sites contain different atoms. This leads to the formation of families of related structures as indicated below.

Diamond, C^{4-t}	Body Centred Cubic, Y^{8-c}	Cubic Close Packed, Cu^{12-cc}
↓	↓	↓
sphalerite, $Zn^{4-t}S^{4-t}$	CsCl, $Cs^{8-c}Cl^{8-c}$	AuCu, $Au^{12-co}Cu^{12-co}$
	MgAg, $Mg^{8-c}Ag^{8-c}$	
↓	↓	↓
fumatinite, $Cu_3^{4-t}Sb^{4-t}S_4^{4-t}$	Fe_3Al, $Fe_3^{8-c}Al^{8-c}$	AuCu3, $Au^{12-co}Cu_3^{12-co}$
	Fe_3Si, $Fe_3^{8-c}Si^{8-c}$	
	↓	
	Li_2AgSb, $Li_2^{8-c}Ag^{8-c}Sb^{8-c}$	

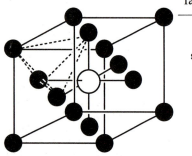

A diagram showing an octahedral hole in a ccp arrangement of atoms.

In the closest packed structures the equivalent spheres occupy 74.05% of the available space. The remaining volume represents empty interstitial sites which have specific geometric characteristics. In particular there are as many octahedral sites as there are atoms (A) in the close packed structures which can accommodate spheres (X) with a radius of 0.414, i.e. ($\sqrt{2}-1$), of the A atoms in the compound AX. In addition, there are twice as many tetrahedral cavities with a radius of 0.225, i.e. ($\frac{1}{2}\sqrt{6} - 1$), of the A atoms. If all the tetrahedral cavities are occupied the compound has the stoichiometry AX_2 (see *Radius ratios*).

Therefore, it is possible to describe the structures of binary compounds AX_n in terms of close packed arrangements of the host atom, A, with specified occupations of the interstitial sites by X. The arrangements of octahedral and tetrahedral sites in ccp are shown opposite. For example, occupation of all the octahedral sites in a ccp host lattice leads to the sodium chloride structure. Since the six octahedral sites are themselves arranged around each atom in an octahedral manner both sodium and chloride ions have octahedral co-ordination geometries and the structure may be described in the following economical fashion:

NaCl — cubic close packed lattice of chloride ions with all the octahedral sites occupied by sodium ions — $Na^{6-o}Cl^{6-o}$.

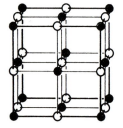

The rock-salt structure based upon a ccp arrangement of chloride ions with all the octahedral holes occupied by sodium ions.

If the metal ion has a radius which exceeds that of the anion, e.g. KF, RbF, and CsF, the metal ions take up the ccp arrangement and the fluoride ions occupy the interstitial sites. This reversal of site occupancies is denoted by the prefix, **anti-**, to the description. Thus, the above examples may be described as anti-NaCl structures. Since both sites are octahedral this is not a profound difference. The use of anti- does assume a greater significance when the alternative sites have different co-ordination numbers.

If all the octahedral sites in an hcp arrangement are filled an alternative structure, described as the **nickel arsenide** structure, results. The centres of the octahedral sites in hcp generate trigonal prisms around the atoms which are close packed. Consequently, although filling the octahedral sites generates a structure of formula AX, the co-ordination geometries of the A and X sites are different. The A atoms have octahedral co-ordination geometries but the close packed X atoms have trigonally prismatic geometries – 6-tp. The nickel arsenide structure may be described as follows:

NiAs – a hexagonal close packed arrangement of arsenic atoms with all the octahedral sites occupied by nickel atoms – $Ni^{6-o}As^{6-tp}$. An anti-NiAs structure would have the less electropositive atom in the octahedral sites.

MX_2 Compounds

If only half the octahedral sites of ccp and hcp structures are filled the stoichiometry becomes MX_2 (or M_2X in the anti-structures) and the ions no longer have identical co-ordination numbers. In a situation where not all the interstitial sites are occupied there are alternative structural possibilities which are best described by specifying the extent to which particular layers are occupied.

Alternation of occupied and empty layers

$CdCl_2$ — cubic close packed arrangement of chloride ions with the cadmium ions occupying octahedral sites in alternant layers — $Cd^{6-o}Cl_2^{3-p}$ (p = pyramidal)

CdI_2 — hexagonal close packed arrangement of iodide ions with the cadmium ions occupying octahedral sites in alternant layers — $Cd^{6-o}I_2^{3-p}$.

These structures are also described as layer structures and are only observed when the anions are polarizable and the metal–anion bonding is partially covalent. The absence of metal ions in alternate layers means that the repulsion between anions in adjacent layers is not compensated for by attractive forces between oppositely charged ions. Consequently these structures are observed usually when the formal charge on the anion is −1.

There are more complex structures where all the planes are partially occupied. For example, Fe_2N has alternate planes which are $2/3$ and $1/3$ of the octahedral sites occupied.

More importantly, the situation where all the planes are half-occupied leads to the following important structural types based upon hcp host lattices:

$CaCl_2$ — found also in $CaBr_2$, $FeO(OH)$, and Co_2C;

α-PbO_2 — found also in the high pressure form of TiO_2;

α-$AlO(OH)$ — found also in α-$FeO(OH)$.

Occupation of tetrahedral sites

The occupation of all the tetrahedral vacancies in a ccp lattice leads to the stoichiometry MX_2. This is known as the **fluorite** structure found in CaF_2 and many other related ionic salts with this stoichiometry. The tetrahedral holes form a cube around each of the atoms which define the close packing so the fluorite structure may be represented by: $Ca^{8-c}F_2^{4-t}$.

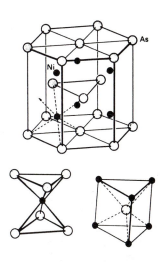

The nickel arsenide structure based upon an hcp arrangement of arsenic atoms with the octahedral holes occupied by nickel atoms. The local geometries of the As (trigonal prismatic) and Ni (octahedral) atoms are shown.

Tetrahedral sites
in ccp

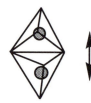

Tetrahedral sites
in hcp

More repulsion

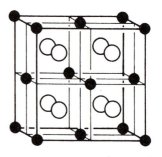

The fluorite structure showing the locations of
the tetrahedral holes in a ccp arrangement.

The **anti-fluorite** structure found in Li_2O for example has the lithium atoms representing close packing and the oxygens in cubic sites; $Li_2^{4-t}O^{8-c}$.

By analogy with NaCl and NiAs a structure based upon hcp anions with all the tetrahedral sites filled is possible. However, this structure is electrostatically unfavourable because neighbouring tetrahedra are face sharing. The resultant electrostatic repulsions between cations occupying the tetrahedra would be excessive:

If only half the tetrahedral sites in the ccp arrangement are occupied several alternative MX structures result:

ZnS — sphalerite – all layers occupied to generate $Zn^{4-t}S^{4-t}$.

PbO — anti-structure of 'close packed' metal atoms with alternant layers of tetrahedral sites occupied by oxygen forming a layer structure – $Pb^{4-spy}O^{4-t}$ (spy = square pyramidal).

PtS anti-structure with layers of 'close packed' metal atoms with sulfur atoms occupying half of the tetrahedral sites – $Pt^{4-sp}S^{4-t}$ (4-sp = square planar).

If only half of the tetrahedral sites of a hcp lattice are occupied the resulting structure is wurtzite, $Zn^{4-t}S^{4-t}$.

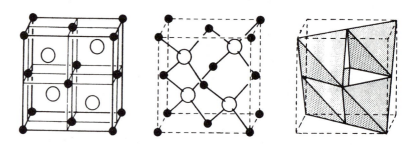

The sphalerite (zinc blende) structure.

Summary of common infinite structural types

Sodium chloride, NaCl
MO (M = Mg, Ca, Sr, Ba, Cd; V, Mn, Co, Ni; Nb, Mo, Ta, Zr;
La, Eu, Ce, Yb, Pu)
MS (M = Mg, Ca, Sr, Ba, Cd, Pb; Mn, Zr; La, Eu, U, Th, Pu)
MX (X = Cl, Br, I; M = Li, Na, K)
MN (M = Sc, Ti, V, Cr, Zr, Hf, Nb, Re; La, Pu)
AgX (X = F, Cl, Br)

Fluorite, CaF_2
MF_2 (M = Ca, Sr, Ba, Cd, Hg, Pb, Hf, Eu)
MO_2 (M = Zr, Bi, Ce, U)
MH_2 (M = Ti, Cr, La, Er, Gd)

Anti-fluorite, M_2X
M_2O (M = Li, Na, K, Rb)
M_2S (M = Li, Na, K)

Nickel arsenide, NiAs
MX (X = S, Se; M = V, Cr, Fe, Ni)
MTe (M = Cr, Mn, Fe, Pd, Pt, U)
MX (X = As, Sb; M = Mn, Fe, Ni)
CoS, CrH

Diamond structure: C, Si, Ge, Grey-Sn

The CsCl structure. The ions at the corners are shared by eight cells and are surrounded by eight nearest neighbour ions.

Zinc blende, sphalerite, ZnS
MS (M = Be, Mn, Zn)
MX (X = Se, Te; M = Zn, Cd, Hg)
MX (X = P, As, Sb; M = Al, Ga, In)
BX (X = P, As)
SiC

Wurtzite, ZnS
MN (M = B, Al, Ga, In)
ZnX (X = O, S, Se)
BeO
MnX (X = S, Se)
SiC

Caesium chloride, CsCl
CsX (X = Cl, Br, I)
TlX (X = Cl, Br)
NH_4Cl

Body centred cubic alloys
NaTl, AgLi, CuZn, CuPd

Rutile, TiO_2
CrX_2 (X = F, Cl)
MF_2 (M = V, Fe, Co, Ni, Cu, Zn, Mg)
MO_2 (M = V, Cr, Mn, Nb, Mo, Ru, Rh, Pd, Ta,
Re, W, Os, Ir, Pt;
Ge, Sn, Pb)
MgH_2

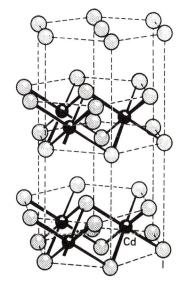

Cadmium chloride, $CdCl_2$
MCl_2 (M = Mg; Mn, Fe, Co, Ni, Cd)
TaS_2

Cadmium iodide, CdI_2
MCl_2 (M = Ti, V)
MX_2 (X = Br, I; M = Mg;
Ti, V, Cr, Fe, Cd)
MX_2 (X = S, Se; M = V,
Zr, Hf, Pt; Sn)

The CdI_2 structure

Anti-cadmium chloride, $M(OH)_2$
(M = Mn, Fe, Co, Ni, Zn, Cd), Cs_2O

D

Example:

Intramolecular
dative bonding

Dative bond (Co-ordinate bond)

See also *Acids and bases* and *Co-ordination compound*.

A dative or co-ordinate bond is a covalent bond which originates from the donation of an electron pair from one atom (or group of atoms) to an empty orbital on a second atom. It may be represented either by an arrow or as a polar covalent bond as follows:

$$R_3N:\rightarrow BX_3 \quad \leftrightarrow \quad R_3N^+\!\!-\!\!^-BX_3$$

The relative contributions of the two *resonance* forms depends on the extent of donation. A dative bond can also occur within the molecule through π-orbitals, e.g. BF_3.

Although the origin of the electron pair in a co-ordinate bond is usually a lone pair on the donating atom, e.g. $:NH_3$, it can also be an electron pair residing in a bonding σ- or π-orbital.

Examples:

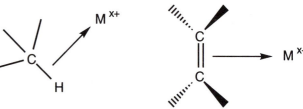

Agostic bond from
C–H to the metal

Metal π-complex

Although most donors are main group ions and molecules, low oxidation co-ordination compounds with filled d-orbitals may also function as σ-or π-donors, as in the examples below.

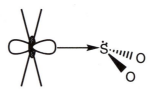

Filled d_{z^2} as sigma donor
in square planar d^8 complex

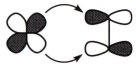

Filled d_{xz} acting
as π-donor

The donating and accepting orbitals must have the same symmetry for the dative bond to form.

Dipole moment

For a +ve charge $+q$ and –ve charge $-q$ separated by a distance r the *dipole moment, $\mu = qr$,* i.e. it is a vector quantity. Molecules act as electric dipoles if the charge distribution in them corresponds to a separation of regions of partial positive and negative charges. This occurs when the vector sum of the bond and lone pair dipole moments have a non-zero resultant. A molecule with a permanent electric dipole is called a polar molecule. A non-polar molecule, although it might be composed of polar bonds, has a net zero electric dipole moment if the dipole vectors cancel.

Examples:

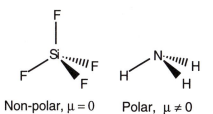

Non-polar, $\mu = 0$ Polar, $\mu \neq 0$

The dipole moment gives useful information regarding the availability of lone pairs, e.g. NH_3, $\mu = 1.47$ D and NF_3, $\mu = 0.23$ D.

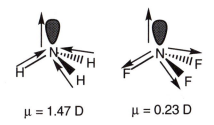

$\mu = 1.47$ D $\mu = 0.23$ D

(N.B. The dipole component is associated with the lone pair in a hybridized orbital.)

Dipole moment measurements can also distinguish between isomers, e.g.:

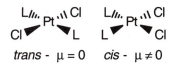

trans - $\mu = 0$ *cis* - $\mu \neq 0$

E

Donating abilities of common ligands

1 electron: F, Cl, Br, I, H, CH_3, NH_2^*, OR^*, SR, NO (bent).

2 electron: O^*, NH_3, OH_2, PR_3, CO, CN^-, CNR, CH_2, C_2H_4, N_2, η–H_2.

3 electron: NO (linear), η-C_3H_3, η-C_3H_5 (allyl), N, CR.

4 electron: η-C_4H_4, η-C_4H_6 (butadiene).

5 electron: η-C_5H_5.

6 electron: η-C_6H_6.

* These ligands are also capable of donating an additional 2 electrons through their π-orbitals
η-refers to the ligands bonded to the metal in a sideways π-fashion.

Main group examples

M = central atom
X = monovalent bonded atom
E = lone pair

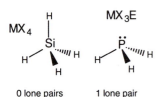

MX_4

0 lone pairs
4 bonding pairs

MX_3E

1 lone pair
3 bonding pairs

MX_2E_2

2 lone pairs
2 bonding pairs

MXE_3

3 lone pairs
1 bonding pair

Effective atomic number rule

(Octet rule, Noble gas rule, Eighteen electron rule)
Applies to molecules where covalent bonding is strong and all the valence orbitals are being used in either homo- and hetero-polar covalent bonds or are occupied by non-bonding electron pairs. The total number of valence electrons around the central atom is therefore 8, if ns and np valence orbitals are available and 18 if nd, $(n+1)s$ and $(n+1)p$ orbitals are available.

The effective atomic number (EAN) is calculated from the following sum:

EAN = no. of valence electrons of the central atom
    + no. of electrons donated by the ligands
    + the negative charge associated with whole ion (zero if the compound is neutral; if the compound has a positive charge subtract the total charge on ion).

Examples of main group molecules and transition metal complexes which obey the rule are shown below. Examples of the electron donating capabilities of ligands are shown in the margin.

Formula	$Cr(CO)_5(PR_3)$	$NiH(PR_3)_4^+$	$Fe(CO)_4^{2-}$
Central atom	6	10	8
Ligands	5x2 (CO)	1 (H)	4x2 (CO)
	2 (PR_3)	4x2 (PR_3)	
Sub total	18	19	16
Charge	0	+1	–2
Electron total	18	18	18

$[ReH_9]^{2-}$	$[W(CN)_8]^{4-}$	$[Mo(CNR)_7]^{2+}$	$Cr(CO)_6$	$Fe(CO)_5$	$Ni(CO)_4$
ML_9	ML_8E	ML_7E_2	ML_6E_3	ML_5E_4	ML_4E_5
0 lone pairs	1	2	3	4	5 lone pairs
9 bond pairs	8	7	6	5	4 bond pairs
tricapped trigonal prism	dodecahedron, square-antiprism*	pentagonal bipyramid, capped-octahedron,* capped trigonal prism*	octahedron	trigonal bi-pyramid, square pyramid*	tetrahedron

*alternative geometries

N.B. The lone pairs are not stereochemically active. They occupy orbitals localized predominantly on the metal and are either involved in 'back donation' to the π^*-orbitals of the ligands or are non-bonding with respect to the ligand σ-orbitals.

If the molecule is unable to achieve the octet or 18-electron configurations by virtue of donation from the ligands then element–element bonds may be formed.

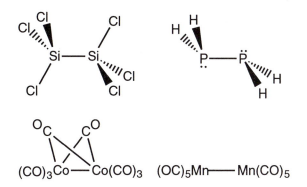

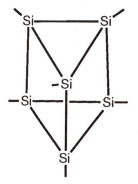

$(CO)_3Co$——$Co(CO)_3$ $(OC)_5Mn$——$Mn(CO)_5$

In these polynuclear compounds the total number of element–element bonds, x, may be easily calculated from the following formulae:

$1/2[8 \times n$ – total no. of valence electrons in molecule] **main group;**
$1/2[18 \times n$ – total no. of valence electrons in molecule] **transition metal.**

Si_6R_6 $R = {}^tBu$

Examples:
$Co_2(CO)_8$ 34 valence electrons, $x = 1/2\,[\,18 \times 2 - 34] = 1$ Co–Co bond
$Fe_3(CO)_{12}$ 48 valence electrons, $x = 1/2\,[18 \times 3 - 48] = 3$ Fe–Fe bonds
Si_6R_6 30 valence electrons, $x = 1/2\,[\,8 \times 6 - 30] = 9$.
The compound has a structure based on a trigonal prism with 9 Si–Si bonds.

Electrode potential

A redox (**red**uction–**ox**idation) reaction consists of the transfer of electrons between the reactants, with the electron donor being described as the **reducing agent** and the electron acceptor as the **oxidizing agent**. All redox reactions may be expressed as the sum of two half-cell reactions which are defined with respect to a standard hydrogen electrode.

$$Mg(s) + Zn^{2+}(aq) \rightleftharpoons Mg^{2+}(aq) + Zn(s)$$

$$Mg(s) \rightleftharpoons Mg^{2+}(aq) + 2e^-$$

$$2e^- + Zn^{2+}(aq) \rightleftharpoons Zn(s)$$

It is conventional to write all half-cell reactions as reduction processes:

$$Mg^{2+}(aq) + 2e^- \rightleftharpoons Mg(s)$$

$$Zn^{2+}(aq) + 2e^- \rightleftharpoons Zn(s)$$

and the processes are abbreviated as $Mg^{2+}(aq)/Mg$ and $Zn^{2+}(aq)/Zn$. They are described as redox couples relative to the standard hydrogen electrode:

$Pt \mid H_2(g) \mid H^+(aq)$ for which the standard reduction potential, $E^{\ominus}$, is zero by convention.

Redox couples under standard conditions may be placed in an electro-chemical activity series such as that shown on page 28. The most strongly oxidizing couples are placed at the top with the most strongly reducing at the bottom.

Ox/Red	$E^{\ominus} N$
$F_2/2F^-$	2.87
Co^{3+}/Co^{2+}	1.82
Au^{3+}/Au	1.50
$Cl_2/2Cl^-$	1.36
VO_2^+/VO^{2+}	1.00
Fe^{3+}/Fe^{2+}	0.77
$I_2/2I^-$	0.54
Cu^{2+}/Cu	0.34
$2H^+/H_2$	0.00
Co^{3+}/Co	− 0.28
Cd^{2+}/Cd	− 0.40
Zn^{2+}/Zn	− 0.76
Mn^{2+}/Mn	− 1.18
Al^{3+}/Al	− 1.66
Na^+/Na	− 2.71
Li^+/Li	− 3.04

Periodic trends

	$E^{\ominus} N$
Li^+/Li	−3.04
Na^+/Na	−2.71
K^+/K	−2.92
Rb^+/Rb	−2.99
Cs^+/Cs	−3.02
$F_2/2F^-$	2.87
$Cl_2/2Cl^-$	1.36
$Br_2/2Br^-$	1.07
$I_2/2I^-$	0.54

N.B. The alkali metals show a limited range of electrode potentials (reducing ability), but the halogens show a larger range of oxidizing ability.

Inorganic chemists use such tables to estimate the *Gibbs free energy changes* associated with a reaction, i.e. whether a redox reaction is thermodynamically favourable. For example, the standard reduction potentials for the $Mg^{2+}(aq)/Mg$ and $Zn^{2+}(aq)/Zn$ couples are −2.36 V and −0.76 V respectively.

For the process:

$$Mg(s) + Zn^{2+}(aq) \rightleftharpoons Mg^{2+}(aq) + Zn(s)$$

the corresponding e.m.f. for the cell based upon these components, $E^{\ominus}$, is given by $-(-2.36) - 0.76 = 1.60$ V.

The sign of the **reduction** potential of the magnesium couple is reversed in the above calculation because magnesium metal is being **oxidized** to magnesium ions. The change in standard Gibbs energy, $\Delta G^{\ominus}$, is given by the equation:

$$\Delta G^{\ominus} = - nFE^{\ominus}$$

where n is the number of electrons associated with the redox reaction and F is the Faraday constant, equal to 96 485 C mol^{-1} (coulombs per mole of electrons).

Even when the number of electrons participating in each half-cell is different the standard reduction potential for the reaction is calculated by using the standard reduction potentials for the half-cells **without any multiplication.**

For the reaction:

$$2Cr^{2+}(aq) + Cu^{2+}(aq) \rightleftharpoons Cu(s) + 2Cr^{3+}(aq)$$

the appropriate half reactions and their standard reduction potentials are:

$$Cr^{3+}(aq) + e^- \rightarrow Cr^{2+}(aq) \qquad E^{\ominus} = -0.41 \text{ V}$$
$$Cu^{2+}(aq) + 2e^- \rightarrow Cu(s) \qquad E^{\ominus} = 0.34 \text{ V}$$

the standard potential for the reaction being given by:

$$E^{\ominus} = 0.34 - (-0.41) = 0.75 \text{ V}$$

because the half-cell potentials represent the potential **per electron.**

If a new half-cell standard reduction potential is calculated from literature values it is necessary to take into account the numbers of electrons participating in the half reactions. In order to eliminate possible errors it is best to carry out the calculation in terms of standard Gibbs energies as in the example:

(1) $Fe^{2+}(aq) + 2e^- \rightarrow Fe(s) \quad E^{\ominus}_1 = -0.44$ V
(2) $Fe^{3+}(aq) + e^- \rightarrow Fe^{2+}(aq) \quad E^{\ominus}_2 = 0.77$ V
(3) $Fe^{3+}(aq) + 3e^- \rightarrow Fe(s) \quad E^{\ominus}_3 = ?$

The calculation of $E^{\ominus}_3$ makes use of the equation $\Delta G^{\ominus} = - nFE^{\ominus}$:

(1) $\Delta G^{\ominus}_1 = -2 \times F(-0.44)$
(2) $\Delta G^{\ominus}_2 = -F(0.77)$
(3) by Hess's law: $\Delta G^{\ominus}_3 = -F(2 \times -0.44 + 0.77) = -F(-0.11)$
so that $E^{\ominus}_3 = -\Delta G^{\ominus}_3/3F = F(-0.11)/3F = -0.037$ V.

Limitations associated with the use of standard electrode potentials

(1) The values of $E^?$ used represent thermodynamic data and do not provide any indications concerning the rate of the reaction.

(2) The data refer to standard conditions and the reaction may be undertaken in a laboratory under non-standard conditions of concentration, temperature, or pressure. The Nernst equation may be used to modify the electrode potential to estimate the effect of these non-standard conditions.

The Nernst equation for the general equilibrium

$$aA + bB \rightleftharpoons cC + dD$$

may be written as

$$E = E^\ominus - \frac{RT}{nF} \ln \frac{[C]^c[D]^d}{[A]^a[B]^b}$$

For a reaction at equilibrium the concentrations of A, B, C, and D; [A], [B], [C], and [D] are equilibrium concentrations and the quotient in the Nernst equation is equal to K_{eq}, with E being equal to zero. This leads to the useful equation:

$$\ln K_{eq} = \frac{nFE^\ominus}{RT}$$

The quotient in the Nernst equation is most commonly affected by changes in pH and complex formation. In particular the complexing abilities of the two oxidation states of a couple are never equal. For example, in the $Fe^{3+}(aq)/Fe^{2+}(aq)$ couple ($E^? = 0.77$ V) ligands such as CN^-, F^-, and NCS^- bind more strongly to $Fe^{3+}(aq)$ than to $Fe^{2+}(aq)$ and therefore stabilize the III state and make Fe^{III} a less effective oxidizing agent.

For example, the dissociation constant for $[Fe(CN)_6]^{4-}$ exceeds that of $[Fe(CN)_6]^{3-}$ by a factor of 10^7 so that, at equilibrium, the ratio $[Fe^{2+}(aq)]/[Fe^{3+}(aq)]$ has a value of 10^7 compared to the conditions in the absence of cyanide ions, thus:

$$E^\ominus[Fe(CN)_6^{3-} / Fe(CN)_6^{4-}] = 0.77 - \frac{RT}{F} \ln \frac{[Fe^{2+}(aq)]}{[Fe^{3+}(aq)]}$$

$$= 0.77 - 0.0592 \log_{10} 10^7 = 0.36 V$$

By contrast ligands such as phen and bipy bind more strongly to Fe(II) than to Fe(III) and in the presence of these ligands Fe(III) behaves as a stronger oxidizing agent.

$$E^?[Fe(phen)_3^{3+}/Fe(phen)_3^{2+}] = 1.12 \text{ V}$$
$$E^?[Fe(bipy)_3^{3+}/Fe(bipy)_3^{2+}] = 0.96 \text{ V}$$

Addition of OH^- to $Fe^{3+}(aq)$ and $Fe^{2+}(aq)$ leads to precipitation of insoluble hydroxides, but the higher oxidation state compound is much less soluble than the one with the lower oxidation state.

The solubility products for the two hydroxides are:

$$Fe(OH)_3 \rightleftharpoons Fe^{3+}(aq) + 3OH^- \qquad K_{sp} = 2.0 \times 10^{-39}$$
$$Fe(OH)_2 \rightleftharpoons Fe^{2+}(aq) + 2OH^- \qquad K_{sp} = 7.9 \times 10^{-16}$$

and suggest that $[Fe^{2+}(aq)]/[Fe^{3+}(aq)]$ has a value of

Activity series

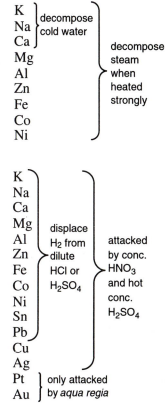

K ⎫ decompose
Na ⎬ cold water
Ca ⎭

Mg   decompose
Al   steam
Zn   when
Fe   heated
Co   strongly
Ni

K
Na
Ca
Mg   displace
Al H_2 from   attacked
Zn   dilute   by conc.
Fe   HCl or HNO_3
Co H_2SO_4 and hot
Ni   conc.
Sn H_2SO_4
Pb

Cu
Ag

Pt ⎫ only attacked
Au ⎭ by *aqua regia*

$7.9 \times 10^{-16}/ 2 \times 10^{-39} = 3.95 \times 10^{-13}$ and if $[OH^-] = 1.0$ then, under these non-standard conditions:

$$E = 0.77 - 0.0592 \log_{10}(3.95 \times 10^{-13}) = 0.036V$$

This leads to the important conclusion that it is thermodynamically more favourable to oxidize $Fe^{2+}(aq)$ to $Fe^{3+}(aq)$ under basic conditions. The stabilization of metal ions in higher oxidation states under basic conditions is quite general.

(3) The standard electrode potentials refer to aqueous solutions and a change in solvent may lead to significant differences in the value of the redox couple. (See *Non-aqueous solvents.*)

Electron affinity

$$A(g) + e^-(g) \rightarrow A^-(g) \quad \Delta H^{\ominus} \quad E_{ea} = - \Delta H^{\ominus}$$

E_{ea}, the electron affinity, is the negative of the electron capture enthalpy, $\Delta H^{\ominus}$, defined above. It is the internal energy *released* when an electron is added to an atom (or indeed a molecule) in its ground state. The values may be either negative or positive. The process is particularly favourable for atoms one electron short of an inert gas configuration, i.e. the enthalpy change is negative. Second and higher electron capture enthalpies are invariably positive, because of the repulsion experienced by the incoming additional electron. Some values are given in the table opposite.

Values of E_{ea} at 298 K in kJ mol^{-1}

O; 136	O$^-$; −850	F; 342
S; 194	S$^-$; −538	Cl; 358
Se; 189		Br; 336
N; 3		I; 308
P; 66		
As; 71		

Electron deficient

An electron deficient molecule has fewer valence electrons involved in covalent bonds than the number of orbitals available. For example, B_2H_6 has 14 valence orbitals but only 12 electrons and therefore conventional two-centre two-electron bonds cannot accurately describe the structure. Almost by definition such molecules are electrophilic and react with nucleophiles unless additional steric or electronic factors intervene. Often the electron deficiency is relieved in an intramolecular fashion by the formation of intramolecular co-ordinate bonds. Examples are given by the bonding in BF_3 (see *Dative bonding*) and that in diborane shown opposite.

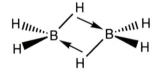

Dative bonds from B–H σ-orbitals

The bonding in B_2H_6 may also be described in terms of three-centre two-electron bonds for the B–H–B bridge bonds (within the molecular orbital framework). The twelve valence electrons then occupy 6 bonding molecular orbitals : 4B–H terminal and 2B–H–B bridge bonds.

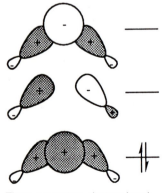

Three-centre two-electron bond

Electronegativity

The electronegativity of an atom is a measure of its power to attract electrons towards it when it is part of a compound. Qualitatively it is an important concept because the electronegativity difference defines the polarity of a bond, but it has not proved possible to define it unambiguously in a quantitative fashion. The Mulliken scale relates the electronegativity coefficient (χ) to the average of the ionization potential (I), and the electron affinity (E_{ea}):

$$\chi = {}^1\!/_2(I + E_{ea})$$

The Pauling scale suggests that the difference between the expected and observed bond energy for a heteroatomic bond may be related to the electronegativity difference:

$$\chi_A - \chi_B = [1/96.49(D_{AB} - 1/2 D_{AA} - 1/2 D_{BB})]^{1/2}$$

D_{AB}, D_{AA}, and D_{BB} are the bond dissociation energies for AB, A$_2$, and B$_2$. Both the Mulliken and Pauling scales are limited by the lack of accurate thermochemical data and have been replaced by the Allred–Rochow scale. The basis of the scale is that the electronegativity of an element is related to the force of attraction experienced by an electron at a distance from the nucleus which is equal to the covalent radius of the particular atom. According to Coulomb's law this force (F) is given by

$$F = Z_{eff} e^2 / r_{cov}{}^2$$

where Z_{eff} is the effective nuclear charge experienced by the electron, r_{cov} is the covalent radius of the atom and e is the unit of electronic charge. Z_{eff} is estimated as $Z - S$ where Z is the atomic number of the atom and S is a shielding factor whose value is calculated by applying Slater's rules. The best fit of the Allred–Rochow values with those on the Mulliken and Pauling scales are given by the equation:

$$\chi = 3590 \, Z_{eff}/r_{cov}{}^2 + 0.744$$

when r_{cov} is expressed in picometres. The χ values are usually rounded off to one decimal place. Some typical values are given in the table opposite.

The periodic table exhibits the general tendencies for the value of χ to increase from left to right across a period and to decrease down each group. Elements with values of $\chi \leq 1.8$ are metals, those with values of $\chi \geq 2.1$ being non-metals. The few elements having values of χ between 1.8-2.1 are known as metalloids (Ge, Te, and At). The polarity or otherwise of a chemical bond depends upon the difference in χ values of the participating elements. If $\Delta\chi < 1.7$ the bond is likely to be covalent with some ionic character. If $\Delta\chi > 1.7$ the bond is likely to be predominantly ionic.

Element	χ
Li	1.0
Be	1.5
B	2.0
C	2.5
N	3.0
O	3.5
F	4.1

Group 17	χ
F	4.1
Cl	2.8
Br	2.7
I	2.2

Group 16	χ
O	3.5
S	2.4
Se	2.5
Te	2.0

Group 15	χ
N	3.1
P	2.1
As	2.2
Sb	1.8
Bi	1.7

Group 14	χ
C	2.5
Si	1.7
Ge	2.0
Sn	1.7
Pb	1.6

Note the large increase in electronegativities across the first long period.

For the post-transition elements the large change in electronegativity between elements of the first two long rows is particularly significant.

Electroneutrality principle

In a co-ordination compound the central ion attempts to attain a net charge lying between (-1) and ($+1$) by electron donation from the ligands and back donation from the metal to the ligands. For example, although the formal charge on cobalt in $[Co(NH_3)_6]^{3+}$ is +3 the transfer of say 0.4e from each NH_3 to the metal which accompanies the formation of the co-ordinate bond reduces the charge to 0.6e. Unfortunately there is no simple way of calculating the extent of electron transfer in a specific complex, because it depends on the oxidation state of the metal and the other ligands present.

Multiple bond formation can play an important role in this charge equalization process.

	$Ni(CO)_4$	OsO_4
Formal oxidation state	0	8+
	Back donation to π^* CO reduces $-$ve charge on Ni resulting from forward donation.	Forward donation from O filled π-orbitals to empty d orbitals to reduce large +ve charge on metal.

The bonding in the chlorate(VII) ion, ClO_4^-, may be regarded as being between a neutral central chlorine atom which participates in three Cl=O double bonds and one single bond to O^-. This and the bonding of the central chlorine atoms with charges of +1, +2, and +3 are shown in the diagram. According to the electroneutrality principle the canonical forms with 2+ and 3+ charged central atoms are less likely to contribute to the bonding of the ion.

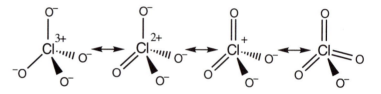

In the ions ClO_4^-, SO_4^{2-}, PO_4^{3-}, and SiO_4^{4-} the double bond character decreases along the series because of the decrease in formal charge on the central atom from +7 to +4.

Electrophile

A reagent which is preferentially attracted towards regions of high electron density in a substrate molecule, e.g. H^+ and NO_2^+.

Electropositivity

Elements which readily lose electrons to form elements in positive oxidation states are described as electropositive, e.g. alkali metals. They have low electronegativity coefficient values.

Entropy

In a molecule not all the energy is stored in chemical bonds, because a molecule is able to rotate, to vibrate, and in the gas or liquid phase also to translate. Therefore, the total energy must express these important components. For example, if a molecule changes phase from solid to liquid to gas the chemical bonds usually do not change dramatically, but the translational, vibrational, and rotational energy terms change markedly. These phase changes are accompanied by entropy changes. Entropy is related to the degree of disorder in a system. If the melting of a solid or the boiling of a liquid occurs reversibly at the transition temperature the entropy change is given by

$$\Delta S = q/T$$

where ΔS is the entropy change, q is the amount of heat absorbed and T is the transition temperature (in degrees K).

The standard entropies of gases are invariably considerably greater than those of liquids, which in turn are greater than those of solids (see table in margin). For chemical reactions the following generalizations usually hold.

element	$S^{\ominus}$ (J K^{-1}mol^{-1})
Au(s)	47.7
Hg(l)	77.4
Rn(g)	176

(1) If a gas is produced in a chemical reaction from solid reactants then $\Delta S^{\ominus}$ is positive, e.g:

$$MCO_3(s) \quad \rightleftharpoons \quad MO(s) \ + \ CO_2(g)$$

M	Mg	Ca	Sr	Ba	Zn	Cd
$\Delta S^{\ominus}$ J K^{-1} mol^{-1}	175	165	171	172	175	176

The virtual constancy of the entropy change is particularly noteworthy.

(2) When monodentate ligands are replaced by chelating ligands $\Delta S^{\ominus}$ is positive because the number of molecules on the right hand side of the equation exceeds that on the left.

$$[Ni(NH_3)_6]^{2+}(aq) \ + \ 3en(aq) \ \rightleftharpoons \ [Ni(en)_3]^{2+}(aq) \ + \ 6NH_3(aq)$$
$$\Delta S^{\ominus} = 84 \text{ J K}^{-1} \text{ mol}^{-1}$$
$$[Ni(NH_3)_6]^{2+}(aq) \ + \ 2dien(aq) \ \rightleftharpoons \ [Ni(dien)_2]^{2+}(aq) \ + \ 6NH_3(aq)$$
$$\Delta S^{\ominus} = 130 \text{ J K}^{-1} \text{ mol}^{-1}$$

en = NH$_2$CH$_2$CH$_2$NH$_2$ – bidentate ligand;
dien = NH(CH$_2$CH$_2$NH$_2$)$_2$ – tridentate ligand.

(3) The solvation of metal ions from the gas phase into aqueous solution leads to an entropy change the magnitude of which depends upon the charge and size of the ion. The more highly charged the ion is the greater is the degree of ordering in the solvated ion and therefore the greater is the loss of translational, rotational, and vibrational freedom by the bound solvent molecules.

$H^+(g) \quad \rightarrow \quad H^+(aq) \qquad \Delta S^{\ominus} = 0$ by convention
$Tl^+(g) \quad \rightarrow \quad Tl^+(aq) \qquad \Delta S^{\ominus} = 125.5 \text{ J K}^{-1} \text{ mol}^{-1}$
$Tl^{3+}(g) \quad \rightarrow \quad Tl^{3+}(aq) \qquad \Delta S^{\ominus} = -192 \text{ J K}^{-1} \text{ mol}^{-1}$

(4) When a solid ionic salt dissolves in water the corresponding entropy change is positive, providing the ions do not order the solvent molecules greatly because of their high charges.

$$MX(s) \quad \rightleftharpoons \quad M^+(aq) \ + \ X^-(aq)$$

MX	KCl	KBr	KI	KNO$_3$	K$_2$SO$_4$
$\Delta S^{\ominus}$ J K^{-1} mol^{-1}	75	87	107	46	46

When the anions or cations are much more highly charged the entropy change is less positive and the solubilities of the salts correspondingly decrease.

From the examples discussed above it is clear that entropy can be related to the degree of order in the system. A positive entropy change for a chemical process is always associated with more degrees of freedom for the products of the reaction, either by virtue of having more product molecules with translational freedom or because the product molecules are more loosely held together. The implications of entropy changes on chemical equilibrium are discussed in the section on *Gibbs energy*.

Exchange energy

A pair of electrons with parallel spins experience less repulsion than a pair with antiparallel spins. The difference is described as the exchange energy, K. This difference results from spin correlation, i.e. the quantum mechanical effect which enhances the probability of two electrons with opposite spins being found close together. The relative exchange energies for alternative electron configurations is estimated by calculating the number of pairs of electrons with parallel spins and multiplying it by K.

For three electrons the alternative configurations shown in the boxes in the margin have exchange energies of $3K$ and K respectively. Therefore the former, with the maximum spin multiplicity is more stable by $2K$. For n parallel spins there are $n(n-1)/2$ pairs of electrons with parallel spins. This is the basis of Hund's Rule which states that the ground state of an atom is that having the greatest spin multiplicity, i.e. that having most unpaired electrons.

The same principles apply to high and low spin transition metal complexes. For example, a d^5 high spin octahedral complex (see diagram opposite) has an exchange energy of $10K$, but only has $4K$ stabilization in the low spin state.

The high spin configuration, although favoured by $6K$ exchange energy, has 2Δ less crystal field stabilization energy. For a strong field such as CN^- where Δ is large a low spin state is favoured since $2\Delta >> 6K$ (see *Ligand Field Theory*).

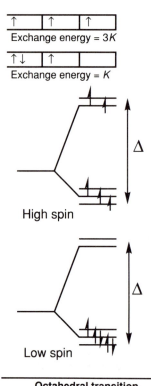

Exchange energy = 3K

Exchange energy = K

High spin

Low spin

Octahedral transition metal complexes

High spin	Low spin
d^4 $t_{2g}^3 e_g^1$	t_{2g}^4
$[MnF_6]^{3-}$	$[Mn(CN)_6]^{3-}$
d^5 $t_{2g}^3 e_g^2$	t_{2g}^5
$[Fe(C_2O_4)_3]^{3-}$	$[Fe(CN)_6]^{3-}$
d^6 $t_{2g}^4 e_g^2$	t_{2g}^6
$[Fe(OH_2)_6]^{2+}$	$[Fe(CN)_6]^{4-}$

Δ: $CN^- > C_2O_4^{2-} > H_2O > F^-$

Exchange energies for octahedral complexes

	High spin	Low spin	Gain in LFSE
d^4	6K	3K	Δ
d^5	10K	4K	2Δ
d^6	10K	6K	2Δ
d^7	11K	9K	Δ

These data suggest that d^6 complexes should form low spin complexes at lower Δ values in the spectrochemical series since the energy difference is $2\Delta - 4K$, cf. d^5 where it is $2\Delta - 6K$.

Gibbs energy

Gibbs energy, G, is sometimes referred to as **free energy** or as **Gibbs free energy**. Gibbs energy is defined as $G = H - TS$, where H = enthalpy, T = absolute temperature and S = entropy. Changes in Gibbs energy, ΔG, are more useful in chemistry because they determine whether reactions are energetically favourable.

The second law of thermodynamics states that in a spontaneous process the change in *entropy* is equal to or greater than zero ($\Delta S \geq 0$) for the system and its surroundings. In the discussion of entropy it was noted that it is important to distinguish in a chemical reaction that energy which is tied up in chemical bonds and that which is distributed amongst the translational, rotational, and vibrational degrees of freedom. Therefore, the total energy change associated with a chemical reaction, 'the change in Gibbs energy'. is given by

$$\Delta G \qquad = \qquad \Delta H \qquad - \qquad T\Delta S$$

| change in Gibbs energy | change in enthalpy | change in entropy x absolute temperature |

The entropy change for a process at constant temperature is defined by the equation

$$\Delta S = q/T$$

where q = heat absorbed by the system at the constant temperature T. For a spontaneous chemical process which leads to an enthalpy change ΔH the corresponding change in the enthalpy of the surroundings is $-\Delta H$. The change in entropy of the surroundings is therefore $-\Delta H/T$. The second law may be written as

$$\Delta S_{\text{system}} + \Delta S_{\text{surroundings}} \geq 0$$

$$\text{or} \quad \Delta S_{\text{system}} - \Delta H/T \geq 0$$

$$\text{or} \quad \Delta H - T\Delta S \leq 0$$

$$\text{i.e.} \quad \Delta G \leq 0$$

Therefore the Gibbs energy change for the reaction, ΔG, may be related directly to the second law of thermodynamics. In particular, if $\Delta G \leq 0$ the reaction is thermodynamically favourable, and will achieve equilibrium defined by the equilibrium constant, K_{eq} (as defined in the margin). The equilibrium constant is related to the standard change in Gibbs energy, $\Delta G^{\ominus}$, for the reaction by the following relationship:

$$\Delta G^{\ominus} = -RT \ln K_{eq} \quad (\text{at } 298 \text{ K} \quad \log_{10} K_{eq} = -0.000175 \times \Delta G^{\ominus})$$

if $\Delta G^{\ominus} = 0$ then $K_{eq} = 1$, and the reaction is in 50:50 equilibrium

if $\Delta G^{\ominus} < 0$ then $K_{eq} > 1$, and the reaction proceeds towards the products (providing there is no kinetic barrier), and the more negative the value of

For the reaction:

$$A + B \rightleftharpoons C + D$$

the equilibrium constant is defined by the equation:

$$K_{eq} = \frac{[C][D]}{[A][B]}$$

$\Delta G^{\ominus}$ the more favourable the equilibrium constant is towards the products, $K_{eq} \gg 1$.

Effect of temperature on ΔG

The relationship, $\Delta G = \Delta H - T\Delta S$, has the following important consequence. If ΔH is positive the reaction equilibrium can be altered to favour the products by increasing the temperature providing that ΔS is positive (see *Entropy* for a discussion of the factors influencing the sign and magnitude of entropy changes). Examples are given below.

For the decomposition reaction:

$$MCO_3(s) \rightleftharpoons MO(s) + CO_2(g)$$

$\Delta H^{\ominus}$ is positive, and $\Delta S^{\ominus}$ is positive (see *Entropy*)

For example $\Delta H^{\ominus}$ ($BaCO_3$) = + 269.3 kJ mol^{-1}

$\Delta S^{\ominus}$ ($BaCO_3$) = 172 J K^{-1} mol^{-1}

$\Delta G = 269300 - T \times 172$

so that $\Delta G = 0$ when $T = 269300/172 = 1565$ K = 1292°C.

The decomposition reaction only becomes favourable at 1292°C when the entropic term outweighs the enthalpic term.

For the solubility reaction: $MX(s) \rightarrow M^{n+}(aq) + X^{n-}(aq)$ $KClO_4$ and $CaSO_4$ have similar $\Delta G^{\ominus}{}_s$ values but the former has a positive value of $\Delta S^{\ominus}{}_s$ and the latter has a negative value of $\Delta S^{\ominus}{}_s$. Therefore, as the temperature increases the former increasingly dissolves in water, whereas the latter becomes less soluble.

More generally the relative contributions to ΔG can be summarized schematically as follows:

ΔG +ve	(1) ΔH +ve, ΔS -ve; ΔG +ve for all T		$K_{eq} \ll 1$
ΔG +ve or −ve	(2) ΔH +ve, ΔS +ve; ΔG − ve at higher T when $T\Delta S > \Delta H$	(3) ΔH −ve, ΔS −ve; ΔG − ve at lower T when $T\Delta S < \Delta H$	K_{eq} depends upon T
ΔG −ve	(4) ΔH −ve, ΔS +ve; ΔG −ve for all T		$K_{eq} \gg 1$

(1) $\frac{1}{2}O_2(g) + H_2O(l) \rightarrow H_2O_2(l)$ $\Delta H^{\ominus}{}_{298} = 98$ kJ mol^{-1}
$T\Delta S^{\ominus}{}_{298} = -21$ kJ mol^{-1}

(2) $NH_4Cl(s) \rightarrow NH_4^+(aq) + Cl^-(aq)$ $\Delta H^{\ominus}{}_{298} = 14.7$ kJ mol^{-1}
$T\Delta S^{\ominus}{}_{298} = 21.4$ kJ mol^{-1}

(3) $2Mg(s) + O_2(g) \rightarrow 2MgO(s)$ $\Delta H^{\ominus}{}_{298} = -1204$ kJ mol^{-1}
$T\Delta S^{\ominus}{}_{298} = -64.4$ kJ mol^{-1}

(4) $2O_3(g) \rightarrow 3O_2(g)$ $\Delta H^{\ominus}{}_{298} = -286$ kJ mol^{-1}
$T\Delta S^{\ominus}{}_{298} = 40$ kJ mol^{-1}

Although reaction (4) is thermodynamically favourable the kinetics of the process are such that the high activation energy of the reaction enables O_3 to exist for sufficient periods of time for its reactions to be studied. (See *Stability*.)

Halogens

The halogens are the group 17 elements of the Periodic Table (F, Cl, Br, I, and At).

H

Hess's law

n	ns^2np^5
2	F
3	Cl
4	Br
5	I
6	At

Hess's law derives from the first law of thermodynamics which states that energy can neither be created nor destroyed, although it may be converted from one form to another. Hess's law states that the overall reaction enthalpy is the sum of the reaction enthalpies of the individual reactions into which a reaction may be theoretically divided.

The individual steps may not be reactions or processes that can be studied in the laboratory and may indeed be hypothetical. The only requirement is that they should be balanced as far as the stoichiometry of the reaction is concerned.

An example of the use of Hess's law is the calculation of the enthalpy of decomposition of calcium carbonate into calcium oxide and carbon dioxide. A direct measure of this quantity would be experimentally difficult, but it may be estimated from the enthalpy changes when calcium carbonate and calcium oxide are separately dissolved in aqueous hydrochloric acid.

$\Delta H_{lattice}$ (kJ mol^{-1})			
NaF	930	AgF	969
NaCl	788	AgCl	912
NaBr	752	AgBr	900
NaI	704	AgI	886

(1) $CaO(s) + 2HCl(aq) \rightarrow Ca^{2+}(aq) + 2Cl^-(aq) + H_2O(l)$

(2) $CaCO_3(s) + 2HCl(aq) \rightarrow Ca^{2+}(aq) + 2Cl^-(aq) + H_2O(l) + CO_2(g)$

If equation (1) is reversed and then added to equation (2) the result is:

$\Delta H_{sol}(M^+) + \Delta H_{sol}(X^-)$			
NaF	929	AgF	991
NaCl	784	AgCl	846
NaBr	753	AgBr	815
NaI	713	AgI	775

(3) $CaCO_3(s) \rightarrow CaO(s) + CO_2(g)$.

Hess's law indicates that $\Delta H(3) = \Delta H(2) - \Delta H(1)$.

The following examples of the use of Hess's law are shown as thermochemical cycles. These and others are of great use in inorganic chemistry for understanding trends in various properties.

The lattice energies of AgX decrease less than anticipated from size effects because of increased contributions from polarization effects.

Data taken from *Inorganic energetics*, W.E.Dasent, Cambridge University Press, 1982, p147.

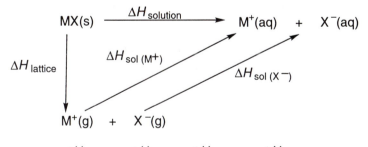

$$\Delta H_{solution} = \Delta H_{lattice} + \Delta H_{sol\ (M^+)} + \Delta H_{sol\ (X^-)}$$

(See *Hydration enthalpy* for definition of Δh_{sol}.)
The above cycle may be used to account for solubility trends, e.g.

NaF < NaCl < NaBr < NaI

AgF > AgCl > AgBr > AgI.

The lower decomposition occurs for the smallest cation because: $\Delta H_{lattice}(MO) - \Delta H_{lattice}(MCO_3)$ is greatest.

For the smallest cation the contraction associated with $CO_3^{2-} \rightarrow O^{2-}$ has a greater impact upon the lattice energy.

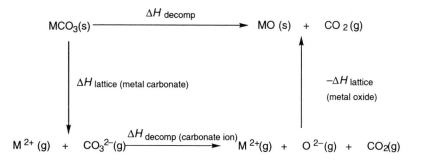

$$\Delta H_{decomp} = -\Delta H_{lattice\ (oxide)} + \Delta H_{decomp\ (carbonate\ ion)} + \Delta H_{lattice\ (metal\ carbonate)}$$

See *Lattice energy* for definition of $\Delta H_{lattice}$, *Ionization energy* for definition of ΔH_{ion}.

The above cycle may be used to account for decomposition temperature trends, e.g:

$$MgCO_3 < CaCO_3 < SrCO_3 < BaCO_3.$$

Couple	$E^{\ominus}/v$
V^{3+}/V^{2+}	−0.26
Cr^{3+}/Cr^{2+}	−0.41
Mn^{3+}/Mn^{2+}	1.49
Fe^{3+}/Fe^{2+}	0.77
Co^{3+}/Co^{2+}	1.77

$\Delta H_{ion(2+)}$ refers to the ionization:
$M^{2+}(g) \rightarrow M^{3+}(g) + e^{-}$

Element	$\Delta H_{ion(3+)}$ (kJ mol^{-1})
V	2829
Cr	2987
Mn	3251
Fe	2957
Co	3231

$$M^{3+}(aq) + e^{-} \xrightarrow{\Delta H_{reduction}} M^{2+}(aq)$$

$-\Delta H_{solv\ (3+)}$ $\Delta H_{Solv\ (2+)}$

$$M^{3+}(g) \xrightarrow{-\Delta H_{ion\ (2+)}} M^{2+}(g)$$

$$\Delta H_{reduction} = -\Delta H_{solv\ (3+)} + \Delta H_{solv\ (2+)} - \Delta H_{ion\ (2+)}$$

The above cycle may be used to account for trends in standard reduction potentials, $E^{\ominus}(M^{3+}/M^{2+})$, examples of which are provided in the table opposite for the first transition series. The $-\Delta H_{ion\ (2+)}$ term is particularly dominant and accounts not only for the general trend of the M^{3+} metal ions becoming progressively more reducing but also the discontinuity at Mn which has a larger third ionization energy, $\Delta H_{ion\ (3+)}$, than simple extrapolation would suggest.

Homoleptic

Compounds containing only one type of ligand are described as homoleptic, e.g. $MoMe_6$, $[ReH_9]^{2-}$, $Mo_2(NMe_2)_6$. If the ligand is monatomic and the complex is uncharged then the compound is described as a binary compound, e.g. $TiCl_4$.

Hund's rules

See Exchange energy.

Hybridization

A hybrid orbital is a linear combination of atomic orbitals centred on a single atom. This mixing of atomic orbitals has the effect of concentrating the electron density of the resultant hybrid into more specific regions of space and shifting the nodal surfaces. The mixing has the net effect of improving the overlap with orbitals of an adjacent atom. These effects are illustrated for some sp^x and spd^x hybrids below.

The % orbital character is obtained by squaring the coefficients of the contributions to the hybrid orbitals:

$0.5\ s + 0.866\ p_z \equiv 1/4\ s + 3/4\ p$ (sp^3 hybrid)

$0.574\ s + 0.819\ p_z \equiv 1/3\ s + 2/3\ p$ (sp^2 hybrid)

$0.707\ s + 0.707\ p_z \equiv 1/2\ s + 1/2\ p$ (sp hybrid)

$0.408\ s + 0.707\ p_z + 0.577\ d_{z^2} \equiv 1/6\ s + 3/6\ p_z + 2/6\ d_{z^2}$ (sp^3d^2 hybrid).

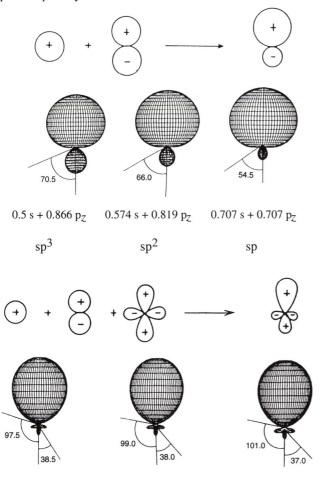

$0.5\ s + 0.866\ p_z$ $0.574\ s + 0.819\ p_z$ $0.707\ s + 0.707\ p_z$

sp^3 sp^2 sp

$0.44s + 0.71p_z + 0.56d_{z^2}$ $0.41s + 0.71p_z + 0.58d_{z^2}$ $0.37s + 0.71p_z + 0.6d_{z^2}$

d^2sp^3

The angles shown on the diagrams are those of the nodal cones.

If n atomic orbitals contribute to the hybridization, n orthogonal (non-overlapping) orbitals are formed, e.g.

      sp hybrids have 2 orthogonal wavefunctions,
      sp^3 hybrids have 4 orthogonal wavefunctions,
      sp^3d^2 hybrids have 6 orthogonal wavefunctions.

These hybrids have their maxima pointing towards the vertices of polygons and three dimensional polyhedra as summarized in the following table.

Some hybridization schemes

Co-ordination number	Arrangement	Composition
2	**Linear**	**sp**, pd, sd
	Angular	sd
3	**Trigonal planar**	**sp^2**, p^{2}d
	Trigonal pyramidal	pd^2
4	**Tetrahedral**	**sp^3**, sd^3
	Square planar	p^{2}d^2, *sp^{2}d*
5	**Trigonal bipyramidal**	sp^{3}d, spd^3
	Square pyramidal	**sp^{2}d^2**, sd^4, pd^4, p^{3}d^2
	Pentagonal planar	p^{2}d^3
6	**Octahedral**	sp^{3}d^2
	Trigonal prismatic	spd^4, pd^5
7	**Pentagonal bipyramidal**	sp^{3}d^3
8	Square antiprismatic	sp^{3}d^4
	Dodecahedral	sp^{3}d^4
9	Tricapped trigonal prismatic	sp^{3}d^5

Those in bold type are the more commonly utilized hybridization schemes. It is noteworthy that as the proportion of p orbital character is increased the angle between hybrids decreases. For main group atoms the predominance of the s and p valence orbitals and the relative unavailability of d orbitals simplifies the choice of hybridization schemes to sp (linear), sp^2 (trigonal), and sp^3 (tetrahedral). Hypervalent compounds such as PF$_5$ (trigonal bipyramid) and SF$_6$ (octahedral) present a problem because although sp^{3}d and sp^{3}d^2 hybridization schemes are geometrically feasible, they are not energetically possible because the d orbitals by virtue of their high energies cannot make as large a contribution to the hybrids as the s and p orbitals. These problems can be circumvented by partial hybridization schemes. For example, PF$_5$ and IF$_7$ may be described as shown opposite.

For transition metals s, p, and d valence orbitals are available for hybrid orbital formation and therefore the sp^{x}d^y hybrids noted in the table represent a reasonably accurate picture of the bonding especially for higher co-ordination numbers.

The hybrid orbitals described above may be used to generate σ-bonding frameworks through the sharing of electrons in covalent bond formation, to act as acceptor orbitals in a dative bond or to accommodate lone pairs of electrons. For example, ammonia may be described initially in terms of four sp^3 hybridized orbitals pointing towards the vertices of a tetrahedron with three hybrids used for N–H covalent bonds and the fourth to accommodate the lone pair. An equally valid starting point would have been the trigonal planar structure based on sp^2 hybrids with the lone pair occupying the p$_z$ orbital.

It is apparent that bending the hydrogens back increases the s character in the lone pair orbital. This is energetically favourable since in the nitrogen atom the 2s valence orbital is significantly more stable than 2p. Indeed the bond angle in ammonia of 107° is even less than the 109.5° predicted for

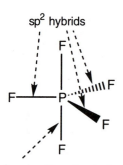

sp^2 hybrids

p$_z$ forming 3-centre 4-electron bonds with axial fluorine atoms

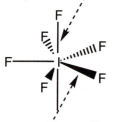

sp hybrid axial bonds

p$_x$p$_y$ multicentred bonds in pentagonal plane

idealized sp^3 hybrids. *Ab initio* calculations have suggested that this orbital has approximately 10% s character, 88% p$_z$ character, and 2% is delocalized on the hydrogen atoms.

   In general, filled non-bonding orbitals will favour a geometry which maximizes the character of its most stable valence orbital in the atom itself, and empty non-bonding orbitals favour the geometry which maximizes the character of the less stable valence orbital.

   BF$_3$ is planar (sp^2 hybridized) because the empty orbital is a pure p-orbital. A pyramidal geometry based on sp^3 hybrids would have introduced more s character into the empty orbital.

   For transition metals the $nd>(n+1)s>(n+1)p$ stability order leads to lone pairs occupying d-orbitals which are stereochemically inactive. For example, in Cr(CO)$_6$ the d^6 electrons occupy d$_{xz}$, d$_{yz}$, and d$_{xy}$ orbitals which are non-bonding as far as the σ-framework is concerned. They do have the correct symmetry to π-bond to the ligands, however. (See *Lone pair.*)

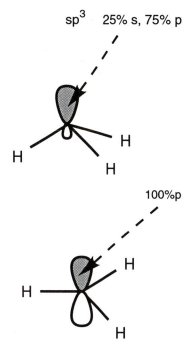

Hydration enthalpy

The hydration enthalpy is the standard enthalpy change for the formation of a hydrated ion from its gaseous state:

$$M^{n+}(g) \quad \rightarrow \quad M^{n+}(aq) \qquad \Delta H^{\ominus} \text{ hydration}$$

$$X^{n-}(g) \quad \rightarrow \quad X^{n-}(aq) \qquad \Delta H^{\ominus} \text{ hydration}$$

   The combined hydration enthalpy of cations and anions of an ionic compound may be calculated from its enthalpy of solution and its lattice energy by using Hess's law. The total hydration enthalpy of a compound may be apportioned into the relative contributions for the individual cations and anions. The procedure leads to the standard ion hydration enthalpies shown opposite. Note that the hydration enthalpies become more negative with the decreasing size of the ion and increasing charge;

$$-\Delta H^{\ominus} \text{ hyd} \propto z/r.$$

$\Delta H^{\ominus}$ hydration /kJ mol^{-1}			
Li$^+$	−520	Mg^{2+}	−1920
Na$^+$	−405	Ca^{2+}	−1650
K$^+$	−321	Sr^{2+}	−1480
Rb$^+$	−300	Ba^{2+}	−1360
Cs$^+$	−277	Al^{3+}	−4690
F$^-$	−506	Fe^{3+}	−4430
Cl$^-$	−364	NH$_4^+$	−301
Br$^-$	−337	OH$^-$	−460
I$^-$	−296	H$^+$	−1100

Hydrogen bond

   A–H$\cdots\cdots$ B

When A and B are electronegative atoms and B has lone pairs a hydrogen bond forms between H and B. For atoms like N, O, and F this interaction can vary in strength from 15 to 165 kJ mol^{-1}. For larger atoms it is generally less than 15 kJ mol^{-1}.

The strongest hydrogen bonds occur in symmetrical molecules like HF$_2^-$. Hydrogen bonds are important in deciding the melting and boiling points of inorganic hydrides (see *Intermolecular forces*). In addition the solid state structures reflect the importance of hydrogen bonding. For example, in H$_2$O the hydrogen bonding completes a distorted tetrahedron about oxygen.

Hypervalent

Those molecules where the number of valence electrons involved in σ-bonding exceeds the number of valence orbitals are described as hypervalent. For example, SH_6, which may be viewed as a simplified version of the very stable SF_6 molecule, has 10 valence orbitals and 12 valence electrons. This assumes, of course, that the sulphur 3d orbitals are of too high energy to contribute to bonding. In such molecules the formation of three-centre four-electron bonds between *trans* hydrogen atoms and the sulphur p orbitals resolves this problem within the *molecular orbital* framework. Strong hydrogen bonds in ions such as $F\text{--}H\text{--}F^-$ may also be described in terms of three-centre four-electron bonds of this type.

In valence bond theory three-centre four-electron bonds are described by the following *resonance* forms:

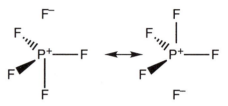

which are particularly favoured for electronegative atoms because of ionic contributions. In molecular orbital theory the bonding, non-bonding, and anti-bonding orbitals (shown opposite) define the three-centre molecular orbitals. Since both the bonding and anti-bonding orbitals have major contributions from the outer atoms the electron density is localized preferentially on these atoms. Therefore, electronegative atoms at these positions favour the occurrence of three-centre four-electron bonds.

anti-bonding

non-bonding

bonding

Inertness and lability

The ability to isolate molecules depends not only on their thermodynamic stability, but also the activation energies for decomposition pathways (see *Stability*). The terms *inert* and *labile* are useful for describing the relative rates of such reactions. Many compounds are thermodynamically unstable either with respect to oxidation or decomposition to the elements, but none the less exist in bottles on the shelf for years. Such *inert* (kinetically stable) compounds have high activation energies associated with their possible decomposition pathways. *Labile* compounds by contrast have low energy pathways and therefore react rapidly to form the thermodynamically favoured products. These qualitative ideas may be placed on a more quantitative basis using half lives derived from kinetic data.

Examples:

$$[Ni(CN)_4]^{2-} \ + \ 4CN^{*-} \ \rightleftharpoons \ [Ni(CN^*)_4]^{2-} \ + \ 4CN^-$$
Labile: half life ~ 30s

$$[Cr(CN)_6]^{3-} \ + \ 6CN^{*-} \ \rightleftharpoons \ [Cr(CN^*)_6]^{3-} \ + \ 6CN^-$$
Inert: half life ~ 24 days

These data suggest that the inertness and lability of co-ordination compounds depends on the metal ion and its *electron configuration*. Octahedral complexes with d^3, d^8, and d^6 (low spin) configurations are inert, those with d^4 and d^9 configurations are very labile.

Element	Valence shell	b.p./K
He	$1s^2$	4
Ne	$2s^2 2p^6$	27
Ar	$3s^2 3p^6$	87
Kr	$4s^2 4p^6$	120
Xe	$5s^2 5p^6$	166
Rn	$6s^2 6p^6$	208

Inert gases

The inert or noble gases are the elements of group 18 of the periodic table: He, Ne, Ar, Kr, Xe, and Rn. (See *Periodic table.*)

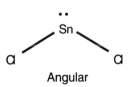
Pyramidal

Inert pair effect

Inappropriately named effect which has nothing to do with the kinetics of the formation of these compounds. It refers to the fact that the post-transition elements have a tendency to form compounds not only with the oxidation state associated with the ionization of all the s and p valence electrons (+*n*), but also the (+*n*–2) oxidation state.

Examples:
Pb^{IV} and Pb^{II}; Sn^{IV} and Sn^{II}; Tl^{III} and Tl^{I}; In^{III} and In^{I}.

Inorganic reactions

The following represents a brief classification of the important classes of inorganic reactions.

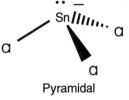
Angular

It is significant that the lone pair in the lower oxidation state remains stereochemically active and is not *localized* in core-like s orbitals. Consequently in both $SnCl_3^-$ and $SnCl_2$ the lone pairs are stereochemically active.

Acid–base reactions

1. Ligand addition reactions

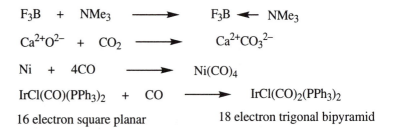

$$F_3B \ + \ NMe_3 \longrightarrow \ F_3B \longleftarrow NMe_3$$

$$Ca^{2+}O^{2-} \ + \ CO_2 \longrightarrow \ Ca^{2+}CO_3^{2-}$$

$$Ni \ + \ 4CO \longrightarrow \ Ni(CO)_4$$

$$IrCl(CO)(PPh_3)_2 \ + \ CO \longrightarrow \ IrCl(CO)_2(PPh_3)_2$$

16 electron square planar   18 electron trigonal bipyramid

2. Ligand elimination reactions

Reverse of the above generating an electron deficient centre.

$$CaCO_3 \ \xrightarrow{\ Heat\ } \ CaO \ + \ CO_2$$

$$Ni(CO)_4 \ \xrightarrow{\ 60°\ C\ } \ Ni \ + \ 4CO$$

Mond process for nickel purification

Dehydration processes:

$$CuSO_4.5H_2O \ \xrightarrow{\ Heat\ } \ CuSO_4 \ + \ 5H_2O$$
$$\text{blue} \qquad\qquad\qquad \text{colourless}$$

3. Ligand extraction/transfer

$$SF_4 \ + \ BF_3 \longrightarrow \ SF_3^+ \ + \ BF_4^-$$
$$\text{Stronger Lewis acid}$$

In this example it is a fluoride ion which is transferred, but similar processes involving other atoms are possible:

$$Me_3NO \ + \ Mo(CO)_6 \longrightarrow \ Mo(CO)_5 \ + \ CO_2 \ + Me_3N$$

Reacts further with solvent
or another ligand

$$IrR_3 \ + \ Me_3NO \longrightarrow \ OIrR_3 \ + \ Me_3N$$
$$(R = \text{bulky alkyl})$$

$$Ca^{2+}O^{2-} \ + \ SiO_2 \longrightarrow \ Ca^{2+}[SiO_3]^{2-}$$

$$6Li_2O \ + \ P_4O_{10} \longrightarrow \ 4Li_3[PO_4]$$

This process may lead to autoionization as in the following example:

$$PCl_5 \ + \ PCl_5 \ \rightleftharpoons \ PCl_4^+PCl_6^-$$

IF_3, BrF_3, and IF_5 undergo similar autoionization processes. The requirement is that the central atom forms cationic (e.g. PCl_4^+ (P^V), neutral PCl_5 (P^V) and anionic PCl_6^- (P^V)) complexes in the same oxidation state.

$$N_2O_5 \rightleftharpoons NO_2^+ \ NO_3^-$$

provides a related example where NO_2^+ behaves as a Lewis acid towards a lone pair on the oxygen atom of NO_3^- in N_2O_5.

4. Ligand replacement reactions

$$F_3BOEt_2 \ + \ Me_3N \longrightarrow F_3BNMe_3 \ + \ Et_2O$$

$$PCl_3 \ + \ AsF_3 \longrightarrow PF_3 \ + \ AsCl_3$$

The relative position of the equilibrium in such reactions depends on:

(a) relative strengths of the dative bonds;
(b) differences in solvation energies
(c) steric effects;
(d) chelate effects.

The larger average bond enthalpy of the P–F bond provides the primary driving force for the reaction.

Hydrolysis reactions
Hydrolysis reactions provide an important sub-class of this reaction where water is the incoming ligand:

$$SiCl_4 \ + \ 2H_2O \longrightarrow SiO_2 \ + \ 4HCl$$

This occurs through intermediate hydrated complexes which polymerize to give the insoluble SiO_2 which has an infinite three-dimensional structure.

For metal ions in +3 and +4 oxidation states hydrated chloride complexes in the solid state hydrolyse when heated to give oxides:

$$2AlCl_3.6H_2O \xrightarrow{Heat} Al_2O_3 \ + \ 6HCl \ + \ 9H_2O$$

$$SnCl_4.6H_2O \xrightarrow{Heat} SnO_2 \ + \ 4HCl \ + \ 4H_2O$$

In some reactions the ligand replacement leads to an interesting sequence of precipitation and redissolution processes:

$$NiCl_2(aq) + NH_3(aq) \rightarrow \underset{\text{insoluble}}{Ni(OH)_2 \downarrow} \ \xrightarrow{NH_3(aq)} \ \underset{\text{soluble}}{[Ni(NH_3)_6]^{2+}}$$

Neutralization reactions

Acid	+	Base	$\rightarrow$	Salt	+	Water
HCl	+	NaOH	$\rightarrow$	NaCl	+	H_2O
NiO	+	$2HNO_3$	$\rightarrow$	$Ni(NO_3)_2$	+	H_2O

5. Metathetical reactions

These reactions involve the formation of an insoluble compound which precipitates from the solution (see solubility for some generalizations concerning soluble and insoluble salts in inorganic solutions).

All chlorides react with concentrated sulfuric acid with evolution of HCl.

Ionic nitrides (e.g. Mg_3N_2) react with acids to give ammonia.

Ionic borides react with acids to give boranes B_nH_{n+4} and B_nH_{n+6}

Ionic carbides reacts with mineral acids to give hydrocarbons and in particular C_2H_2.

$$Ca(NO_3)_2 + Na_2CO_3 \rightarrow 2NaNO_3 + CaCO_3 \quad \text{aqueous solution}$$
$$ZnCl_2 + Na_2S \rightarrow 2NaCl + ZnS \quad \text{aqueous solution}$$
$$TiCl_4 + LiMe \rightarrow TiMe_4 + 4LiCl \quad \text{organic solvent}$$
$$\text{soluble} \qquad \text{soluble} \qquad \text{soluble} \qquad \text{insoluble}$$

There are similar metathetical reactions involving gas evolution. All carbonates, sulfites and sulfides for example react with acids to evolve CO_2, SO_2, and H_2S respectively.

Redox reactions

1. Electron transfer reactions

Reduction-oxidation reactions occur when the formal oxidation states of the atoms in the reactants change during the reaction. The thermodynamic aspects of these reactions are discussed under *electrode potentials*. The rate of transfer of the electron from one ion to a second in a redox reaction can vary over a wide range and does not necessarily proceed by a simple 'hopping' of an electron from one molecule to a second. In the following examples, the first reaction takes place by an electron transfer but in the second there is the additional transfer of a chloride ion:

$$[Fe(CN)_6]^{4-} + [Mo(CN)_8]^{3-} \rightarrow [Fe(CN)_6]^{3-} + [Mo(CN)_8]^{4-}$$

$$[Co(NH_3)_5Cl]^{2+} + [Cr(H_2O)_6]^{2+} \rightarrow [Cr(H_2O)_5Cl]^{2+} + [Co(H_2O)_6]^{2+} + 5NH_3$$

2. Disproportionation reactions

If in a reaction an element in a species simultaneously undergoes oxidation and reduction the process is described as disproportion. For example:

$$Hg_2^{2+} + 4I^- \rightleftharpoons Hg^0 + [HgI_4]^{2-}$$

$$Cl_{2(g)} + 2OH^-_{(aq)} \rightleftharpoons Cl^-_{(aq)} + ClO^-_{(aq)} + H_2O$$

3. Oxidative-addition reactions

Typical X–Y molecules in oxidative addition reactions involving transition metals include :

F_2, Cl_2, Br_2, and I_2; H_2; NCS–SCN; RS–SR.

$Ph_3C–CPh_3$; NC–CN; HCl, HBr, HI; H_2S, $H–SiR_3$

CH_3I, Ph–I, R_3SnCl, $Ph_3PAu–Cl$. These add to low oxidation state transition metal complexes such as $[IrCl(CO)(PPh_3)_2]$ and

$[RhCl(PPh_3)_3]$ (d^8 square-planar) and $[Pt(PPh_3)_3]$ (d^{10} trigonal planar).

The compound adds X–Y to form new M–X and M–Y bonds, observed for transition and post-transition elements which have stable oxidation states n, $n+2$, and sometimes $n+4$, because the addition of the molecules X–Y to ML_n requires the participation of a lone pair on M.

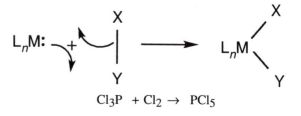

$$Cl_3P + Cl_2 \rightarrow PCl_5$$

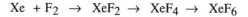

$$Xe + F_2 \rightarrow XeF_2 \rightarrow XeF_4 \rightarrow XeF_6$$

4. Reductive elimination reactions

These are the reverse of the above reactions

$$PBr_5 \xrightarrow{\text{Heat}} PBr_3 + Br_2$$

Polymerization–depolymerization reactions

Thermally induced:

$$N_2O_4 \rightleftharpoons 2NO_2$$
100% solid at −11°C   100% gas at 140°C

$$\text{cyclo-}S_8 \xrightarrow{150°C} -S_8- \longrightarrow S_x \text{ polymer}$$

$$2H_3PO_4 \xrightarrow{\text{Heat}} H_4P_2O_7 + H_2O$$

$$N_2S_2 \xrightarrow{>10°C} \{NS\}_x \text{ polymer}$$

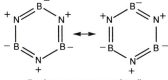

(hydrogen atoms omitted)

pH controlled:

$$2CrO_4^{2-} + 2H^+ \rightleftharpoons Cr_2O_7^{2-} + 2H_2O$$

$$[Fe(H_2O)_6]^{3+} \rightleftharpoons [Fe(H_2O)_5(OH)]^{2+} + H^+$$
pale purple   yellow

$$\longrightarrow [Fe(H_2O)_4(OH)_2]^+ \longrightarrow Fe_2O_3.xH_2O \text{ (rust)}$$
                          insoluble

Hydrolysis followed by polymerization:

$$SiCl_2Me_2 + 2OH^- \rightarrow Si(OH)_2Me_2 \rightarrow [SiOMe_2]_n \text{ (silicone polymer)}$$

Migratory reactions

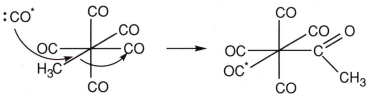

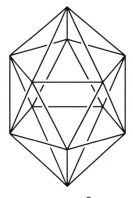

$$B_{12}H_{12}^{2-}$$

Electrophilic and nucleophilic substitution

Electrophilic and nucleophilic substitution reactions at aromatic rings and cages. $B_3N_3H_6$(borazine) which is isoelectronic with benzene undergoes nucleophilic substitution reactions at the boron atoms and addition occurs across the double bonds readily to form saturated derivatives e.g. $B_3N_3H_6Br_6$. $B_{12}H_{12}^{2-}$ and the isoelectronic carborane $C_2B_{10}H_{12}$ which have icosahedral cage structures undergo electrophilic substitution reactions on the cage. Sandwich compounds such as ferrocene $Fe(\eta\text{-}C_5H_5)_2$ (shown in the margin) undergo electrophilic substitution reactions on the cyclopentadienyl rings.

Tautomerism

The dynamic interconversion of isomeric forms is described as tautomerism, e.g:

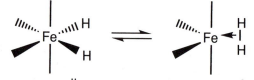

Dihydride Fe^{II} complex   Dihydrogen Fe^0 complex

Intermolecular forces

The forces between molecules are important in deciding their melting and boiling points. There are attractive and repulsive contributions and they are collectively described as van der Waals forces.

Repulsive forces are only significant at a very short range and result from the repulsion of electron clouds when they are forced to overlap significantly, i.e. when the distances between the atoms are shorter than the sum of the van der Waals radii.

Attractive forces and their dependence on the intermolecular separation are summarized below.

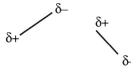

dipole–dipole interaction

London dispersion forces:
induced dipole – induced dipole;
dipole – induced dipole

Interaction	Radial dependence of potential energy	Approximate magnitude (kJ mol^{-1})
Hydrogen bonding	Directional, therefore r dependence is variable	>20 for F, O, N <20 for heavier atoms
Dipole–dipole	$1/r^6$	~ 0.5
London dispersion	$1/r^6$	~ 2.0

Where hydrogen bonding is significant it tends to dominate the attractive interactions and for first row elements can make an enormous difference to boiling points.

	H_2O	CH_3OH	$(CH_3)_2O$
b.p. (°C)	100	64.7	−23.7
m.p.(°C)	0	−97.8	−138.4

If hydrogen bonding is not important then molecules with dipole moments generally have higher boiling points than those which do not.
Example:

	SF_4(polar)	SiF_4 (non-polar)
b.p. (°C)	−38	−86

The importance of hydrogen bonding for the second row atoms is indicated in the following graph of the boiling points of the simplest hydrides of the elements of groups 14, 15, 16, and 17. The abnormally high values for water, ammonia, and HF indicate the effects of hydrogen bonding.

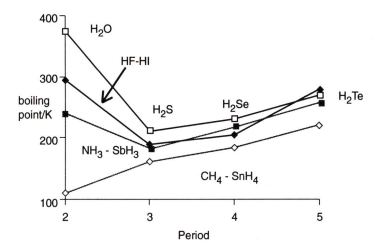

If the molecule does not have a permanent dipole then the London dispersion forces dominate and these depend primarily on the polarizability of the molecule.

Examples:

	F_2	Cl_2	Br_2	I_2
b.p.(°C)	−188	−35	60	185
m.p. (°C)	−219	−101	−7.3	114

	He	Ne	Ar	Kr	Xe	Rn
b.p. (K)	4	27	87	120	166	211

In the following series of molecules the dipole moment is decreasing but the London dispersion forces increase. The latter effect clearly dominates:

	PF_3	AsF_3	SbF_3
b.p. (°C)	−102	63	292

	ClF_5	BrF_5	IF_5
b.p.(°C)	−103	−60	10

	SiH_4	SiF_4	$SiCl_4$
b.p. (°C)	−112	−86	−58

The high volatility and stability of covalent fluorides are important for their applications for isotope separation in the nuclear industry (UF_6) and as anaesthetics and refrigerant gases (CFCs).

If the molecules begin to interact in the solid through covalent bonding the m.p.s and b.p.s begin to change dramatically.

Monomeric		Weakly polymeric		Infinite lattices	
	b.p.°C		m.p.°C		m.p.°C
CO	−192	SO_3	45	SiO_2	2230
CO_2	−79	P_4O_6	175	B_2O_3	2300
CS_2	−46	P_4O_{10}	300(subl)		

In a column of the periodic table when compounds with identical formulae are compared then there is an increased tendency for the compound to be polymeric as one descends the group.

Ionic bond formation

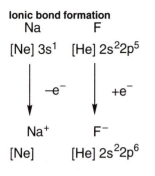

Both complete inert gas configurations

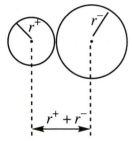

Ionic bond

An ionic bond involves the transfer of electrons from one atom of low electronegativity to a second atom of high electronegativity. The subsequent Coulomb attraction between the ions formed provides a strong driving force for this electron transfer. It depends on the relative sizes (see diagram opposite) of the ions and their charges (see *Lattice energies*). According to Coulomb's Law the attractive energy between charges z^+ and z^- is given by:

$$E \propto z^+z^-/(r^++r^-)$$

Since the *ionization energies* required to form successively more highly charged cations increase rapidly their formation is favourable only when z^+ has the values 1–3. Similar considerations based on the *electron affinities* of the anions limit the formation of anions to values of z^- to 1 or 2. The maximum Coulomb interactions occur when cation and anion are small.

In the solid state ionic molecules generally condense to form infinite structures where the cations and anions are surrounded by 4–8 oppositely charged ions. This increases the Coulomb interactions proportionately.

The $\Delta H_f^{\ominus}$ formation of an ionic compound in the solid state therefore depends on the thermodynamic cycle shown below.

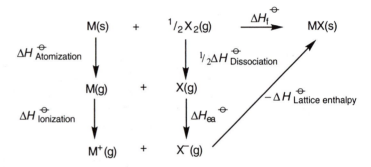

The lattice enthalpy is sufficiently large in an ionic compound to compensate for the large energy inputs involved in getting atoms into the gas phase and forming the resultant ions. There is a continuum of bond types from covalent to ionic depending on the electronegativity difference between the atoms. As the covalent bonding contribution diminishes the bond weakening is generally more than compensated for by the increased ionic contribution.

Ionization energy

Also referred to as ionization potential. The first ionization energy, I, is the minimum energy required for the following process at $T = 0K$:

$$A(g) \rightarrow A^+(g) + e^-(g) \qquad \Delta U = I$$

i.e. it is the internal energy required to remove completely an electron from the gaseous atom (or molecule) in its ground state. For use in calculations as $\Delta H_{ionization}$, 6 kJ mol^{-1} (the value of RT at 298°C) should be added to the value of I. Subsequent ionizations are similarly defined, although the process of ionization becomes progressively more difficult. The first ionization energy of an atom is very sensitive to the effective nuclear charge, the size of the atom, and the repulsions between electrons within the orbitals from which the ionization is occurring. The following general trends are important:

1. *I* generally decreases down a column of the periodic table (increasing size of atom).There are exceptions particularly after filling of d and f shells for the first time.
2. For a horizontal row of the periodic table *I* generally increases with atomic number, but the effects are influenced by the filling of different sub-shells. As the atomic number increases the effective nuclear charge increases because the additional electron does not screen the valence electrons perfectly.
3. Electron repulsion effects facilitate the ionization of electrons from orbitals that are more than half filled.

Isoelectronic molecules

Molecules with the same number of valence electrons are described as isoelectronic. This concept is particularly useful when the resulting molecules have similar properties.

Examples: N_2 CO NO^+ CN^-

All form similar co-ordination compounds because they are all π-acid ligands, although of course the ease with which they co-ordinate differs markedly. If the analogy is stretched too far the similarity is lost, e.g. B–F is *isoelectronic* with N_2 but the multiple bond character is reduced by the polarity $B^{2-}\equiv F^{2+}$. In fact BF is only known at very high temperatures in the gas phase.

BF_3, NO_3^-, CO_3^{2-}, BO_3^{3-}, F_2CO, and FNO_2 are all trigonal planar and isoelectronic.

C_6H_6, $B_3N_3H_6$, $H_3B_3O_3$, and $C_3N_3H_3$ all have planar hexagonal structures but do not exhibit the same degree of aromatic properties as benzene.

Other examples of isoelectronic molecules:

OsO_4 and $OsNO_3^-$

$Fe(CO)_2(NO)_2$, $Co(CO)_3NO$, and $Ni(CO)_4$.

diamond, graphite, and the two structurally similar forms of boron nitride, $(BN)_n$.

P_4 and Ge_4^{4-} in K_4Ge_4 are both tetrahedral clusters.

$B_6H_6^{2-}$ and $C_2B_4H_6$ both have octahedral cluster geometries.

The isoelectronic concept is a powerful one in inorganic chemistry and the noble gas compounds may have been discovered twenty years earlier if more

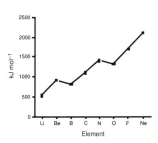

Note the Increase across a period and discontinuities at ns^2np^1 (new shell) and ns^2np^4 (electrons pairing in *p* level).

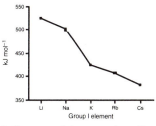

Note the general decrease down the periodic column.

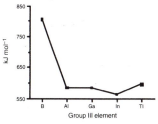

Note the overall decrease down a periodic column, but also discontinuities after filling 3d shell (Ga) and 4f shell (Tl).

attention had been paid to the isoelectronic relationship between $[IF_4]^-$ and $[XeF_4]$

The chemical properties of isoelectronic molecules can be significantly different if they have different numbers of lone pairs or empty orbitals. The following species for example, become progressively more basic:

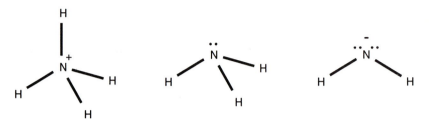

Isomerism

Compounds with the same formula but with different physical (colour, melting point, solubility) or chemical properties are described as isomers. Structural isomerism arises from a different bonding arrangement of atoms and takes the following forms.

Alternative co-ordination polyhedra

$[NiCl_2(PBzPh_2)_2]$ crystallizes with half the molecules adopting a tetrahedral arrangement and the other half a square-planar arrangement. Similar structural manifestations in cage compounds are described as skeletal isomerism.

Co-ordination sphere isomerism

The compound with the empirical formula $[CrCl_3(H_2O)_6]$ occurs as the isomers:

$[Cr(H_2O)_6]^{3+} Cl^-_3$ violet

$[Cr(H_2O)_5Cl]^{2+}Cl^-_2.H_2O$ green; 1 water molecule of crystallization

$[Cr(H_2O)_4Cl_2]^+Cl^-.2H_2O$ green; 2 water molecules of crystallization

Geometric isomerism

Cis and *trans* isomers of square-planar and octahedral compounds are common. Examples:

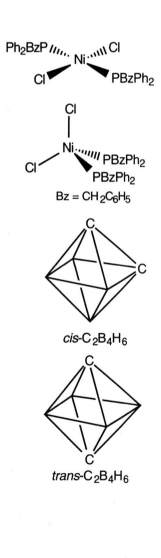

Similar possibilities occur for cage compounds, for example $[C_2B_4H_6]$ which has an octahedral geometry with *cis* and *trans* isomers.

Ionization isomerism

Ionization isomers possess different combinations of ligands in the co-ordination sphere and as counter ion. They give rise to different ions when they dissociate in solution, e.g.

$$[Co(NH_3)_5Br]SO_4 \qquad [Co(NH_3)_5SO_4]Br$$
$$\text{violet} \qquad\qquad\quad \text{red}$$

Linkage isomerism

Ambidentate ligands may bond to a metal atom through different atoms, both of which have lone pairs.

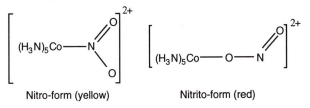

Nitro-form (yellow)   Nitrito-form (red)

Optical isomers

If the mirror images of a molecule are non-superimposable then they form optical isomers. The most commonly encountered examples occur for the octahedral complexes $[Co(en)_3]^{3+}$ (shown below) and cis-$[CoCl_2(en)_2]^+$, en $= H_2NCH_2CH_2NH_2$.

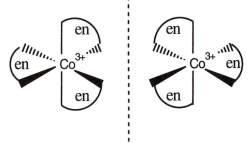

Polymerization isomerism

Monomers and polymers of various lengths may have the same empirical formulae, e.g: monomer: $[Pt(NH_3)_2Cl_2]$; polymer: $[Pt(NH_3)_4][PtCl_4]$.

L

lanthanide	4f	5d
Ce	2	0
Pr	3	0
Nd	4	0
Pm	5	0
Sm	6	0
Eu	7	0
Gd	7	1
Tb	9	0
Dy	10	0
Ho	11	0
Er	12	0
Tm	13	0
Yb	14	0
Lu	14	1

Trends in lattice energies (see *Born –Haber cycle, Ionic bond*)

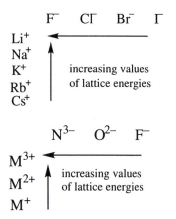

Lanthanides

Or rare earth elements, are those elements from cerium (58) to lutetium (71), inclusive. (See *Periodic table*.)

Lattice energy

The lattice energy is the energy required to convert one mole of a solid ionic compound in its usual lattice structure at absolute zero into its gaseous ions. Theoretically according to the ionic model the calculated lattice energy is given by the Born–Landé equation:

$$U = \frac{NMz^+z^-e^2}{4\pi\varepsilon_o r}\left(1-\frac{1}{n}\right)$$

where N is Avogadro's number, M is the Madelung constant for the particular structure, z^+ and z^- are the positive values of the charges on cation and anion respectively, ε_0 is the permittivity of a vacuum, r = minimum cation-anion separation ($r^+ + r^-$) and n is the Born exponent (in general n has an approximate value of 9). The Born exponent takes into account the repulsions between the filled shells of electrons between oppositely charged ions at short distances. The Madelung constant depends on the geometric arrangement of ions in the lattice and sums the attractions and repulsions which any ion experiences from the other ions in the infinite crystal. From its viewpoint this ion sees spherical shells of increasing radii containing ions with increasing numbers of ions either with the same or opposite charges. Some calculated values of the Madelung constant are given in the table below, together with values of M/v (where v is the number of ions in a formula unit of the compound) and the average coordination numbers for each structure.

Structure	M	M/v	Average co-ordination number
CsCl (8:8)	1.763	0.88	8
NaCl(6:6)	1.748	0.87	6
Wurzite(4:4)		0.82	4
Fluorite(8:4)	2.519	0.84	$5^1/_3$
Corundum((6:4)	4.172	0.83	4.8

The observation that the M/v values are reasonably constant leads to the Kapustinskii equation:

$$U = \frac{1.214\times10^5\, vz^+z^-}{r}\left(1 - \frac{34.5}{r}\right)$$

The Kapustinskii equation is useful for semi-quantitative estimates of lattice energies for interpreting solubility and stability properties of ionic compounds, because it emphasizes the importance of the sizes of ions.

where v is the number of ions in a formula unit of the compound, the Madelung constant which is specific to the lattice type has been replaced by an average value which is no longer lattice dependent. The values of r^+ and r^- are the ionic radii expressed in picometres, $r = r^+ + r^-$.

Ligand

See also Dative bond.

Ligands are ions or neutral molecules which can form co-ordinate or dative bonds with metal centres. They possess lone pairs of electrons and may be classified as σ or π donors. A σ-donor ligand is capable of donating an electron pair only through σ-symmetry orbitals, e.g. H^-, NH_3. A π-donor ligand is capable of donating electron pairs through π-orbitals in addition to σ donation. The π-donor ability increases with negative charge: $F^- < O^{2-} < N^{3-}$. Ligands such as NR_2^- have the ability to donate in only one plane. π-acceptor ligands are capable of donating electron pairs through σ-orbitals and accepting through π-orbitals, e.g. CO, CN^-, N_2, NO^+. Some ligands can accept preferentially in one plane because they possess only one π^* orbital, e.g. SO_2, C_2H_4.

Ambidentate ligand

Ligand capable of donating electron pairs from more than one atomic site. Example: NCS^- in M–NCS and M–SCN. Different metal–ligand geometries occur as shown in the margin.

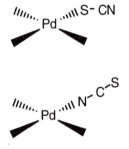

Monodentate ligand

Donation from one lone pair on the ligand to the acceptor site. Example: NH_3 in $[Co(NH_3)_6]^{3+}$.

Polydentate ligand

Donation from several lone pair sites on the ligands to the acceptor site. A ligand which binds to a metal in this way is described as a chelating ligand and may form 4-, 5-, or 6-membered rings as in the examples shown below.

Bidentate ligands

Four-membered rings:

Dithiocarbamates   Xanthates   Carboxylates

Five-membered rings:

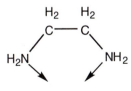

2,2'-bipyridine (bipy)   1,2-bis(dimethylarseno)benzene (diars)   1,2-ethanediamine (en)

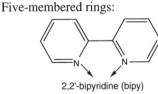

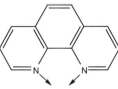

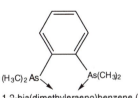

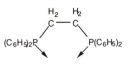

1,10-phenanthroline (phen)   1,2-ethanediylbis(diphenylphosphine) (diphos)

Six-membered rings:

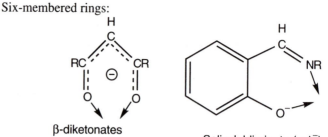

β-diketonates
e.g. acetylacetonate (acac)
2,4–pentadionato–

Salicylaldiminato (sal⁻)

Tridentate ligands

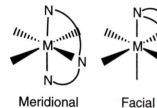

Meridional   Facial

These are of two types; those which are obliged to be planar and those which are flexible, e.g.$H_2NCH_2CH_2NHCH_2CH_2NH_2$ diethylenetriamine (dien), and $(CH_3)_2As(CH_2)_3AsCH_3(CH_2)_3As(CH_3)_2$, bis(3 dimethylarsinylpropyl) methylarsine (triars). The flexible tridentate ligands can co-ordinate to a metal centre in two ways; meridional or facial, as shown in the diagrams opposite. Two examples of obligate planar tridentate ligands are shown below.

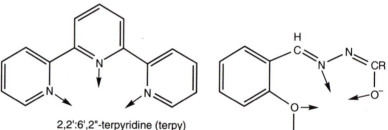

2,2':6',2"-terpyridine (terpy)

Acylhydrazones of salicylaldehyde

Quadridentate ligands

There are three types of quadridentate ligands;

(1) open-chain unbranched molecules, e.g. triethylenetetramine (trien) $H_2N(CH_2)_2NH(CH_2)_2NH(CH_2)_2NH_2$

(2) macrocyclic molecules which are essentially planar and include Schiff bases, porphyrins and phthalocyanins which are shown below.

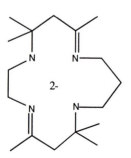

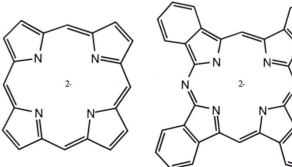

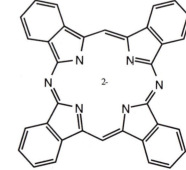

(3) tripod ligands of the type X(—Y)₃ where X = N, P or As and the Y groups are R₂N, R₂P, R₂As, RS, or RSe, with connecting chains (—) between the X and Y groups which are (CH₂)₂, (CH₂)₃, or o-phenylene. Some common examples are shown in the margin.

N(CH₂CH₂NH₂)₃ (tren)
N[CH₂CH₂N(CH₃)₂]₃ (Me₆tren)
N[CH₂CH₂P(C₆H₅)₂]₃ (TPN)
P[o-C₆H₄P(C₆H₅)₂]₃ (QP)
N(CH₂CH₂SCH₃)₃ (TSN)
As[o-C₆H₄As(C₆H₅)₂]₃ (QAS)

Tripod ligands are used to favour the formation of trigonally bipyramidal complexes but they do not always give this result. [Ni(TPN)I]⁺ is a trigonal bipyramid but [Co(TPN)I]⁺ is a square pyramid as is shown in the margin.

Pentadentate and hexadentate ligands

Ethylenediaminetetraacetic acid [EDTAH₄] acts as a hexadentate ligand when the acidic protons are fully dissociated as shown below. If this ion is singly protonated the ion is then a potential pentadentate ligand.

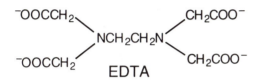

EDTA

Macrocyclic ligands

Some large ring compounds have sufficient flexibility to allow three or more donor atoms to bind to a metal ion. The donor atoms are most commonly nitrogen atoms, though oxygen and sulfur atoms (or a mixed set) also occur. Some examples are shown below. Crown ethers, thia-crown ethers, and aza-thia-crown ethers form complexes with ions of groups 1 and 2. Bicyclic molecules, such as the one shown below, also form stable complexes with ions of groups 1 and 2 elements, the metal ions being in the centre of a cage. The resulting complexes are called cryptates, the ligands being known as cryptands.

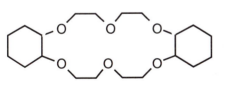

Crown ether   Thia-crown ether

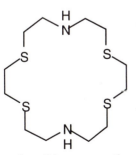

Aza-thia-crown ether   Cryptand

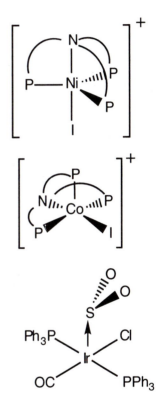

Rare example of a metal acting as a donor.

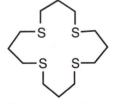

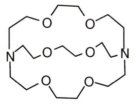

Octahedral
complex

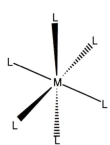

In crystal field
theory

Ligands
represented
by point charges

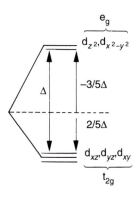

Ligand field theory

This theory has been developed to describe the electronic spectra and the magnetic, thermochemical, and kinetic properties of transition metal complexes. In its original form '*Crystal field theory*' was based on the electrostatic interactions between point charges representing the lone pairs of ligands and the d-orbitals of a transition metal ion. Subsequently it was modified to take into account the potential covalent character of the metal–ligand bonds, M–L, hence the name 'Ligand field theory'.

The electrons in the five d-orbitals illustrated below do not experience the same degree of repulsion from the point charges representing the ligands in an octahedral complex. Electrons in the metal $d_{x^2-y^2}$ and d_{z^2} orbitals which point directly at the ligands experience more repulsion than those in the d_{xz}, d_{yz} and d_{xy} orbitals which have their lobes lying between the x, y, and z axes.

Symmetry arguments show that these differential interactions lead to the energy splitting diagram illustrated in the margin.

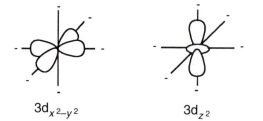

$$3d_{x^2-y^2} \qquad 3d_{z^2}$$

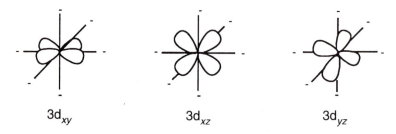

$$3d_{xy} \qquad 3d_{xz} \qquad 3d_{yz}$$

Specifically the $d_{x^2-y^2}$ and the d_{z^2} orbitals experience the same degree of repulsion and are described as doubly degenerate orbitals (e_g). These lie Δ above the set of triply degenerate t_{2g} (d_{xz}, d_{yz}, d_{xy}) orbitals. If one defines a weighted zero point on the energy scale (a barycentre) the energies of these orbitals relative to it are $+2/5\Delta(t_{2g})$ and $-3/5\Delta(e_g)$.

For transition metals the energy gap Δ is comparable to the electron–electron repulsion energy and therefore the usual *Aufbau principle* cannot be applied blindly. Specifically a metal complex with four d-electrons may take up *low spin* $(t_{2g})^4$ or *high spin* $(t_{2g})^3(e_g)^1$ configurations, with only the latter conforming to the Aufbau principle. The relative *crystal field stabilization energies* (CFSE) are:

Low spin $(t_{2g})^4$ $2/5\,\Delta \times 4$ $= 8/5\Delta$
High spin $(t_{2g})^3 e_g$ $2/5\Delta \times 3 - 3/5\Delta$ $= 3/5\Delta$

The low spin configuration has a CFSE Δ larger than the high spin configuration because to form it an electron has been removed from e_g (which lies Δ above t_{2g}) and placed in t_{2g}. However, this transfer has resulted in increased electron–electron repulsion since the additional electron which initially was localized along the x, y, and z axes now resides in the t_{2g} set and is localized in the space between the axes, which it has to share with the other three electrons. The additional electron repulsion energy associated with moving the electron from e_g to t_{2g} is described as the *spin pairing energy*, P. A major contribution to P is the difference in *exchange energy*.

Therefore, the distinction between the high spin and low spin complexes depends on the following inequalities:

High spin	Low spin
Octahedral transition metal complexes	
$d^4\ t_{2g}{}^3 e_g{}^1$	$t_{2g}{}^4$
$[MnF_6]^{3-}$	$[Mn(CN)_6]^{3-}$
$d^5\ t_{2g}{}^3 e_g{}^2$	$t_{2g}{}^5$
$[Fe(C_2O_4)_3]^{3-}$	$[Fe(CN)_6]^{3-}$
$d^6\ t_{2g}{}^4 e_g{}^2$	$t_{2g}{}^6$
$[Fe(OH_2)_6]^{2+}$	$[Fe(CN)_6]^{4-}$

$$\Delta > P \text{ — strong field case, low spin preferred;}$$
$$\Delta < P \text{ — weak field case, high spin preferred.}$$

Since high spin and low spin configurations have different total numbers of unpaired spins ($(t_{2g})^4$ – total spin S=1; $(t_{2g})^3(e_g)$ – total spin S=2) they may be distinguished by their paramagnetic moments, which are proportional to $\sqrt{(4S(S+1))} = [n(n + 2)]^{1/2}$ (see *Magnetism*).

If high spin and low spin configurations differ by two electrons, then the corresponding difference, e.g. $(t_{2g})^5$ (low spin) and $(t_{2g})^3(e_g)^2$ (high spin), in crystal field stabilization energies is 2Δ, because two electrons are demoted from e_g to t_{2g}. The pairing energy is correspondingly increased. (See *Exchange energy*.)

The strength of the ligand field, Δ, depends on the ligands, L, and the following series has been established experimentally:

$I^- < Br^- < S^{2-} < \underline{S}CN^- < Cl^- < NO_3{}^- < F^- < C_2O_4{}^{2-} < H_2O < \underline{N}CS^- < NH_3 < en < bipy < phen < NO_2{}^- < PPh_3 < CN^- < CO.$

underlined atoms are donors

This series cannot be interpreted simply in terms of electrostatic effects, particularly since such neutral ligands such as CO and NH_3 appear higher in the spectrochemical series than small anionic ligands such as F^-. For this reason it is necessary to modify the model to take into account covalency effects. The strength of covalent bonding in such complexes depends primarily on the *electronegativity* of the donor atom (the lower the electronegativity the more efficient the electron pair donation in the co-ordinate M–L bond), and the π-bonding character of the ligand. The latter is important because although the M–L σ-covalent bond occurs through the metal d_{z^2} and $d_{x^2-y^2}$ orbitals, the metal d_{xz}, d_{yz}, and d_{xy} orbitals have the correct symmetry properties to overlap the π-orbitals of the ligands. The spectrochemical series therefore represents a superposition of the following ligand characteristics:

Electronegativity trends:

$$
\begin{array}{ccccccc}
F & < & O & < & N & < & C \\
Cl & < & S & < & P & < & Si
\end{array}
$$

π-bonding character:

π-donors	no π-character	π-acids
F^-, O^{2-}	NH_3, H^-	CO, CN^-
Cl^-, S^{2-}	en	phen, bipy

The ligand positions in the tetrahedron away from the *x*, *y*, and *z* axes lead to a reversal of the d orbital splittings.

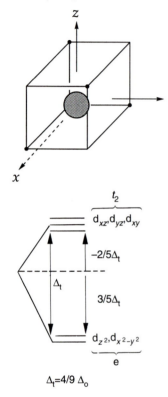

$\Delta_t = 4/9\,\Delta_o$

Ligands with large Δs are good σ-donors and π-acceptors. Those with small Δs are poor σ-donors and good π-donors.

Covalency effects also determine the role of the central metal atom in the spectrochemical series. Specifically, the extent of covalent overlap with the ligand orbitals increases as one descends the periodic table 3d<4d<5d and with the increase in the metal oxidation state. The following trends in Δ can therefore be interpreted:

$$Mn^{2+} < V^{2+} < Co^{2+} < Fe^{2+} < Ni^{2+}$$
$$Fe^{3+} < Co^{3+} < Ir^{3+} < Pt^{4+}$$

Finally the magnitude of Δ depends on the number of ligands (n). For example, the splitting in a tetrahedral complex is $^4/_9\Delta_{octahedral}$ because there are $^2/_3$ the number of ligands and Δ is proportional to n^2. Furthermore, the different geometries of the ligands affect the relative ordering of orbitals. In a tetrahedral complex it is the d_{xz}, d_{yz}, and d_{xy} orbitals which experience more repulsion than $d_{x^2-y^2}$ and d_{z^2} because the former are closer to the ligands than the latter. Therefore the energy level ordering is reversed.

Lone pair

Classically the lone pair is represented by a pair of dots on the atom. However, in modern quantum mechanical descriptions it takes many forms and may be defined as *an electron pair which is occupying an orbital which is effectively non-bonding by virtue of its nodal planes*.

Lone pair orbitals are capable of donating electron pairs to empty acceptor orbitals with matching symmetry properties (see *Dative bond*). The availability of these lone pairs can be markedly influenced by the other substituents on the atom.

If these orbitals are empty then they can function as acceptor orbitals towards ligands with filled donor orbitals with matching symmetry characteristics.

Examples:

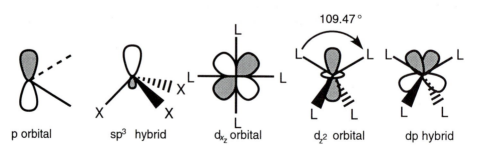

p orbital   sp³ hybrid d_{xz} orbital d_{z^2} orbital   dp hybrid

Magnetism

Diamagnetism

Magnetic property of a material as a result of the circulation of closed (complete) shells of electrons. The magnetism produced acts in opposition to that of the applied field. This is characteristic of molecules with no unpaired electrons.

n	μ_{eff}
1	1.73
2	2.83
3	3.87
4	4.90
5	5.92

Paramagnetism

Magnetic property arising from the unbalanced spin of electrons around the nuclei in molecules possessing unpaired electrons. When a magnetic field is applied the atoms tend to orientate themselves in the direction of the field and possess a positive magnetic susceptibility, i.e. are attracted towards the magnetic field. Paramagnetic effects are generally much larger than diamagnetic effects. Examples; NO_2, NO (1 unpaired electron), O_2 (2 unpaired electrons), and many transition metal ions which can adopt alternative spin states depending on geometry and the ligands, e.g. Ni^{2+} square planar are diamagnetic; tetrahedral are paramagnetic (see *Ligand field theory*). Calculated magnetic moment (spin only formula) $\mu_{eff} = [n(n+2)]^{1/2}$ where n = no. of unpaired electrons (see margin for examples).

Examples of octahedral transition metal complexes with high or low spin configurations:

High spin	Low spin
$d^4\ t_{2g}^3 e_g^1$	t_{2g}^4
$[MnF_6]^{3-}$	$[Mn(CN)_6]^{3-}$
$d^5\ t_{2g}^3 e_g^2$	t_{2g}^5
$[Fe(C_2O_4)_3]^{3-}$	$[Fe(CN)_6]^{3-}$
$d^6\ t_{2g}^4 e_g^2$	t_{2g}^6
$[Fe(OH_2)_6]^{2+}$	$[Fe(CN)_6]^{4-}$

Metals, metalloids, and non-metals

Metals are characterized by the following properties:

(a) they are good conductors of heat and electricity,

(b) they are ductile (capable of being drawn into wire) and malleable (capable of being beaten into sheets), and

(c) they are good reflectors of light and most have a silvery white appearance because they reflect all wavelengths in the visible region. Gold and copper absorb some light in the blue region which causes their characteristic colours.

The melting points of solid metals vary over a tremendous range from Cs (29°C) to W (3380°C) indicating a very wide range in the strength of metal–metal bonds. Mercury (m.p. −39°C) is unique in being a liquid at standard temperature.

Metals are generally insoluble in water, although the more reactive elements (group 1 metals in particular) react with water with the liberation of H_2. Liquid mercury dissolves many metals to form amalgams; a Ag–Hg–Sn amalgam is used for dental fillings. The transition from metallic to non-metallic behaviour occurs along a diagonal line in the periodic table when arranged as shown below.

Metals		*Metalloids*				*Non-metals*	
Li	Be	B	C	N	O	F	Ne
Na	Mg	Al	Si	P	S	Cl	Ar
K	Ca	Ga	Ge	As	Se	Br	Kr
Rb	Sr	In	Sn	Sb	Te	I	Xe
Cs	Ba	Tl	Pb	Bi	Po	At	Rn

The elements enclosed by the bold lines are described as metalloids because it is difficult to define them unambiguously as either metals or non-metals. For example, they are semiconductors with electrical conductivities intermediate between metals and insulators.

The non-metals do not conduct electricity and several of them are gases at the standard temperature (H_2, F_2, Cl_2, O_2, N_2, He, Ne, Ar, Kr, Xe, and Rn). Only one is a liquid (Br_2). The elements C, P, S, and I are the most common solid non-metals and display many of the chemical properties typical of this class of elements.

Metallic bond

As in other areas of chemistry the bonding in metals is not described by one theory alone. Free electron, *molecular orbital,* and *valence bond* concepts have all been adapted to describe various aspects of metallic character and structure. It is important to recognize that the interactions between atoms in metals is in principle no different from that in other infinite structures which are insulators but differs in practice because each metal is surrounded by more nearest neighbours (8–12 generally) and there are insufficient electrons to form localized two-centre two-electron bonds to each of the adjacent metal atoms. Within the *molecular orbital* description this electron deficiency is overcome by forming highly delocalized orbitals which extend over the whole solid and which are very closely spaced in energy. The resultant molecular orbitals are described as bands and the high electrical and thermal conductivities of metals are associated with the partial filling of these bands of molecular orbitals. The most stable orbitals within a band have few nodes and are therefore strongly metal–metal bonding. Progression up the band, however, leads to more nodes and less strong bonding until one reaches the top of the band where there are nodes between each of the atoms and the orbital is strongly antibonding. Therefore, the binding energy of the metal depends on the number of electrons which are introduced into the band using the *Aufbau principle.* In particular, maximum bonding is achieved when the band is half filled and tapers to zero binding for empty or completely filled bands. This model therefore accounts for the changes in binding energies of the second and third row transition metals shown in the margin. The pronounced double humped behaviour of the binding energies of the first transition series results from the non-Aufbau filling of the bands in the centre of the series where the *exchange energies* of the metal atoms are large.

The widths of the bands depend on the overlap of the orbitals in much the same way that the energy difference between bonding and antibonding molecular orbitals in a diatomic molecule. Consequently in a transition metal the d band is relatively narrow because the 3d orbitals are relatively contracted, whereas the 4s and 4p bands are wide because these orbitals are relatively diffuse and overlap well. This difference is represented qualitatively by diagrams such as that shown on page 63.

Within the *valence bond* framework the bonding in metals is described by hybrid orbitals. However, since it is not possible to form simultaneously hybrid orbitals to each of the adjacent metal atoms then it is necessary to

standard enthalpy of atomization (kJ/mol)

K - Zn

Group number

standard enthalpy of atomization (kJ/mol)

Rb - Cd

Group number

standard enthalpy of atomization (kJ/mol)

Cs - Hg

Group number

initially pick out a subset of the metal atoms to which hybrids are formed. Then *resonance* is used to delocalize the wavefunctions over the complete set of atoms.

In the free electron model the initial starting point is that the electrons are free to move like a gas anywhere within the boundaries of the metallic crystal. This problem is therefore analogous to the particle in the box problem and when analysed quantum mechanically leads to quantized energy states which are subject to the Pauli exclusion principle. The effect of the metallic nuclei is then treated as a perturbation on these energy states. This model is most appropriate for the alkali metals where the metallic binding is particularly weak. This free electron model is also particularly useful for accounting for the thermal and electrical properties of metals because it is mathematically relatively easy to calculate the conductivity properties for a system in the absence of strong interactions between the nuclei and the electrons.

Mixed valence compounds

Compounds which apparently have a fractional oxidation state often have the metal ions in two oxidation states within the same compound. For example, Pb_3O_4 is more accurately represented as $2Pb^{2+}Pb^{4+}O_4$. Such compounds are described as Class I mixed valence compounds if the oxidation states are localized and well defined by virtue of having different geometries. Additional examples: $GaCl_2$, GaS, $TlCl_2$, TlS, K_2SbCl_6. The energy required to transfer an electron between the two oxidation states is large, and therefore there is essentially no interaction between the ions. In Class II compounds the compounds have different environments for the different ions, but the sites are sufficiently similar that electron transfer requires energies comparable to the wavelengths found in the visible region. Many of these compounds are highly coloured and are semiconductors, e.g. Eu_3S_4, Na_xWO_3, and $CsAuCl_3$. In Class IIIA delocalization is found within a discrete molecule or ion, e.g. I_3^- or Bi_5^{3+}, and the formal assignment of oxidation states breaks down. In Class IIIB the delocalization extends over the whole lattice and the compounds are generally metallic conductors, e.g. $Ni_{1-x}O$, Ag_2F where the conductivity increases as the degree of non-stoichiometry increases.

Molecular orbital theory

In molecular orbital theory the electron is described by a wavefunction that spreads throughout the molecule and is not localized on a specific atom. Molecular orbitals are constructed by taking a linear combination of atomic orbitals (LCAOs). If there are N valence orbitals contributing to the molecular orbitals they generate N molecular orbitals. The most stable molecular orbital has no nodes in the interatomic bond directions, and the molecular orbitals become progressively less stable as the number of nodes is increased. The electrons occupy the molecular orbitals according to an *Aufbau* procedure. The molecular orbital which is the last one filled in the Aufbau procedure is described as the *highest occupied molecular orbital*

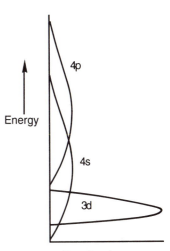

Density of states
The density of states is the number of energy levels per unit energy increment.

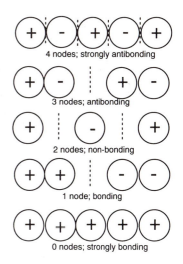

4 nodes; strongly antibonding

3 nodes; antibonding

2 nodes; non-bonding

1 node; bonding

0 nodes; strongly bonding

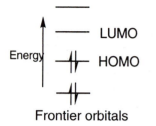

Frontier orbitals

The u and g subscripts refer to the symmetry of the molecular orbitals and in particular to their properties with respect to the centre of inversion which lies at the mid point of the nuclei. The in-phase combination is symmetric with respect to the inversion operation (*gerade*–(g)–even in German) and the out-of-phase combination is asymmetric (*ungerade*–(u)–uneven in German). This molecular orbital analysis is represented by the diagram which represents the energies of the molecular orbitals relative to isolated atoms.

Points at *x, y, z* and -*x, -y, -z* are identical and related by a centre of inversion

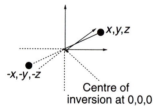

Orbitals which have a centre of inversion are gerade **g** and those which do not are ungerade **u**.

(HOMO) and the first vacant orbital is described as the *lowest unoccupied molecular orbital (LUMO)*. The energies and nodal characteristics of these orbitals are crucial in deciding the reactivities of molecules, and they are called **frontier orbitals**. Molecular orbitals which are more stable than the atomic orbitals in the isolated atom are described as *bonding molecular orbitals,* and those which are less stable as *antibonding molecular orbitals.* (these orbitals are frequently indicated by the superscript[*], e.g. σ^* and π^* representing antibonding σ and π orbitals respectively). If the energy of the molecular orbital is the same as that of the atomic orbitals then it is *non-bonding (NBMO)*.

The simplest example of the application of molecular orbital theory is the hydrogen molecule H_2 since the two atoms each contribute only a single 1s orbital. At the internuclear distance found in the molecule (74 pm) the hydrogen 1s orbitals form two linear combinations of atomic orbitals (LCAOs) which are in-phase and out-of-phase. Both are σ-orbitals because they have cylindrical symmetry about the H–H axis, and the in-phase bonding combination (σ_g) is the more stable because it has no node along the bond direction. The constructive interference of wave functions associated with σ_g leads to an increase in electron density between the nuclei and therefore it is more stable than the 1s orbitals of the isolated hydrogen atoms. The out-of-phase combination (σ_u) results in an orbital which is less stable (antibonding) because of the destructive overlap of wavefunctions between nuclei.

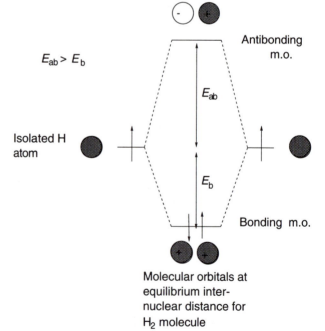

Molecular orbitals at equilibrium internuclear distance for H_2 molecule

Each molecular orbital can accommodate two electrons with opposing spins, so they resemble atomic orbitals and the same *Aufbau* principle can be applied. Each electron which occupies σ_g gains a stabilization energy E_b and those which occupy σ_u are destabilized by E_{ab}. Therefore, the Aufbau principle suggests the following stabilization energies for simple hydrogen-like molecules (electron–electron repulsion energies are ignored):

$$\underset{E_b}{H_2^+\,(\sigma_g^1)} \quad \underset{2E_b}{H_2\,(\sigma_g^2)} \quad \underset{2E_b-E_{ab}}{H_2^-(or\ He_2^+)\,(\sigma_g^2\sigma_u)} \quad \underset{2E_b-2E_{ab}}{H_2^{2-}\,(or\ He_2)\,(\sigma_g^2\sigma_u^2)}$$

Since, $E_{ab}>E_b$ it follows that when two atoms with filled s shells such as H^- or He approach each other the interaction is repulsive and no covalent bond is formed. This leads to the following important conclusion of molecular orbital theory which has a direct analogue in valence bond theory:

Two-orbital two-electron interactions achieve the maximum stabilization energy and correspond to classical bond formation.

Two-orbital four-electron interactions are repulsive and no bond is formed.

If the atoms forming a molecule have s and p valence orbitals the same principle may be used to develop the relevant molecular orbitals for the diatomic molecule. Eight molecular orbitals result and they are illustrated in the figure below:

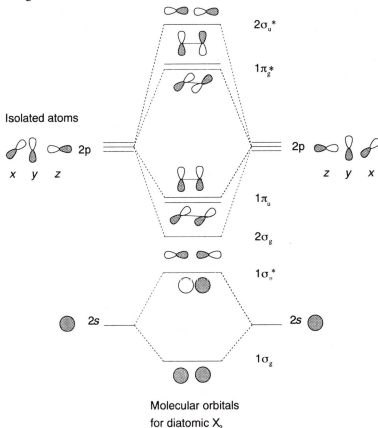

Isolated atoms

Molecular orbitals
for diatomic X_2

The 2s atomic orbitals form bonding and antibonding molecular orbitals $1\sigma_g$ and $1\sigma_u^*$ analogous to those developed above for H_2.

Summary of bond lengths and bond dissociation energies for simple diatomics with 1–4 electrons, i.e. with formal bond orders of 1/2, 1, 1/2, and zero.

	X–X(pm)	D(X–X) (kJ mol^{-1})
H_2^+	106	256
H_2	74	432
He_2^+	108	230
He_2	no covalent bond	

This simplified treatment ignores interactions between 2s and 2p orbitals on the different atoms which can alter the relative energies of $2\sigma_g$ and $1\pi_u$ in diatomics such as B_2, C_2, and N_2.

Summary of bond lengths and bond dissociation energies for some diatomic molecules.

	(X–X) (pm)	D(X–X) (kJ mol^{-1})
B$_2$	159	274
C$_2$	124	603
N$_2$	110	942
O$_2$	121	494
F$_2$	142	155
O$_2$	121	494
S$_2$	189	422
Se$_2$	217	325
Te$_2$	256	261
F$_2$	142	155
Cl$_2$	199	239
Br$_2$	228	190
I$_2$	267	149
N$_2$	110	942
P$_2$	189	477
As$_2$	229	382
Sb$_2$	221	299
Bi$_2$	–	195

The p$_z$ orbitals which point to towards each other also form $2\sigma_g$ and $2\sigma_u^*$ combinations corresponding to in-phase and out-of-phase overlaps. The in-phase and out-of-phase overlap of p$_x$ and p$_y$ orbitals on the two atoms lead to π-bonding and antibonding orbitals π_u and π_g^*. These are described as π-orbitals because they have a node along the internuclear axis (see *Covalent bond*). The in-phase π-bonding combination has an ungerade (u) subscript because although it is bonding it is antisymmetric with respect to the inversion operation. The π-orbitals are doubly *degenerate*, i.e. the p$_x$–p$_x$ and p$_y$–p$_y$ combinations are indistinguishable and therefore have the same energy. The π_u and π_g^* molecular orbitals therefore can each accommodate a total of four electrons, i.e. two pairs of spin paired electrons.

The electronic configurations of diatomic molecules such as N$_2$, O$_2$, and F$_2$ may be derived by applying the *Aufbau principle* to the molecular orbital scheme illustrated in the figure. The N$_2$ molecule has a total of 10 valence electrons leading to the electronic configuration: $(1\sigma_g)^2(1\sigma_u^*)^2 (1\pi_u)^4(2\sigma_g)^2$. The molecule has a formal bond order of three since the $1\sigma_g$ and the $1\sigma_u$ orbitals effectively cancel each other and 6 electrons remain in the bonding molecular orbitals ($2\sigma_g$ and $1\pi_u$) in this simple analysis equal weight is given to σ- and π-orbitals.

The O$_2$ molecule has an additional pair of electrons which must occupy the $1\pi_g^*$ according to the *Aufbau principle*. Since $1\pi_g^*$ is degenerate the electrons enter the orbitals with parallel spins according to Hund's rules, i.e. they minimize electron repulsion by occupying two separate but equal molecular orbitals (see *Exchange energy*). The $1\pi_g^*$ molecular orbital is antibonding and therefore the formal bond order of O$_2$ is reduced to two.

F$_2$ has a pair of electrons which complete the occupation of $1\pi_g^*$ and the formal bond order is reduced to one. The hypothetical Ne$_2$ molecule would be associated with a formal bond order of zero since the $2\sigma_u^*$ orbital is doubly populated. The table in the margin summarizes important bond dissociation energy and bond length data for diatomic molecules and underlines the usefulness of the qualitative molecular orbital ideas developed above. The data also underline the manner in which the strengths of multiple bonds decrease down a group of the periodic table (see *Bond order*). This results primarily from a decrease in the p$_\pi$–p$_\pi$ overlaps.

Nomenclature of inorganic compounds

Inorganic chemistry deals with very different types of compounds and therefore different naming schemes have evolved. Increasingly systematic names based on well defined rules are replacing trivial names which have a long history but do not have the flexibility to be adapted for naming the wide range of compounds that currently exist. These are the more important rules (for a more authoritative discussion see *Nomenclature of inorganic chemistry: Recommendations* (1990) Ed. G.J. Leigh, official IUPAC Publication):

Simple binary names

Carbon disulfide CS_2; boron trifluoride BF_3; copper dichloride $CuCl_2$.

The name of a monoatomic electropositive constituent is simply the unchanged element name. The name of a monoatomic electronegative constituent is the element name with the ending replaced by -ide.

Alternatively the oxidation state of the metal may be used to define the stoichiometry:

MnO_2 manganese(IV) oxide for manganese dioxide;

P_2O_5 phosphorus(V) oxide for diphosphorus pentoxide.

This type of nomenclature can also be extended to ternary, quarternary, etc. compounds:

Phosphorus(V) oxychloride $POCl_3$.

Examples:

Chloride from chlorine
Arsenide from arsenic
Hydride from hydrogen
Oxide from oxygen
Sulfide from sulfur
Carbide from carbon.

Exceptions based on Latin names:

Stannide for tin
Auride for gold
Plumbide for lead
Natride for sodium.

Substitutive names

This is based on organic nomenclature principles and widely used for molecular main group compounds. The organic or inorganic radicals are viewed as substituents on the parent (sometimes hypothetical) hydride:

Dichlorosilane SiH_2Cl_2; chlorodiphenylphosphine PPh_2Cl.

Molecular hydride (or -ane names)

Usually used for the following elements:

B , C , Si , Ge, Sn, Pb, P, As, Sb, Bi, O, S, Se, Te, Po.

Silane SiH_4; germane GeH_4; phosphane PH_3.

In hypervalent compounds the valency is indicated by a superscript added to the Greek letter λ:

PH_5 λ^5–phosphane; SeH_6 λ^6–selenane.

If the hydride is oligomeric then prefix (di, tri, etc.) is added to indicate the number of main group atoms.

Diborane B_2H_6; Decaborane $B_{10}H_{14}$.

Many trivial names still persist, for example ammonia, NH_3; hydrazine N_2H_4.

Cations derived from proton addition to molecular hydrides:

Ammonium $[NH_4]^+$; phosphanium $[PH_4]^+$; hydrazinium $[N_2H_5]^+$.

Replacement names:

carba, aza, thia, oxa may be used to indicate heteroatom substituents, e.g. dicarbapentaborane $B_3C_2H_5$.

Oxoacids and their salts

Based largely on trivial names. Nitric acid HNO_3, -ic ending reserved for higher oxidation state , -ous for lower oxidation state, e.g. HNO_2 nitrous acid. Superlatives are then introduced, e.g. perchloric acid $HClO_4$, hypernitrous acid $H_2N_2O_2$. Salts are usually described by changing the acid name to an -ate ending for the anion, e.g. sodium nitrate $NaNO_3$, potassium sulphate K_2SO_4. Replacement of oxygen by sulfur indicated by the prefix, thio, e.g. thiosulfate, $S_2O_3^{2-}$.

	Systematic name
HNO_3	trioxonitric acid
HNO_2	dioxonitric acid
$HClO_4$	tetraoxochloric acid
$H_2S_2O_3$	trioxothiosulphuric acid
$H_2Cr_2O_7$	μ-oxo-hexaoxodichromic acid (where the μ- indicates a bridging oxygen.)

Condensed acids

Diphosphoric acid $H_3P_2O_7$, sodium triphosphate $Na_3P_3O_9$.

Mixed salts

Magnesium chloride hydroxide $MgCl(OH)$, potassium sodium carbonate $NaKCO_3$.

Additive names

Triphenylphosphine oxide Ph_3PO, ammonia-borontrifluoride $H_3N–BF_3$.

Co-ordination names

This is an additive system primarily developed for inorganic co-ordination compounds which is based on the name of the central atom and the associated ligands.
Neutral complexes:
   Ligands in alphabetical order and then the metal atom. The number of ligands is indicated either by di, tri, tetra, penta, etc., derived from cardinal numerals, or bis, tris, tetrakis, pentakis, derived from ordinals.

Abbreviations for some ligands and their systematic names

More examples

$[Co(NH_3)_6]Cl_3$: hexaaminecobalt(III) chloride.

$K_2[OsCl_5N]$: potassium pentachloronitridoosmate(VI).

$Na[PtBrCl(NO_2)(NH_3)]$: sodium aminebromochloronitrito-*N*-platinate(II)—*N* indicates that the ligand is bonded through nitrogen.

$[Fe(CNCH_3)_6]Br_2$: hexakis(methyl isocyanide)iron(II) bromide.

$K_3[Fe(CN)_6]$: potassium hexacyanoferrate(III).

Abbreviation	Common ligand name	Systematic name
acac(1−)	acetylacetonato	2,4-pentadionato
edta(4−)	ethylene diamine-tetraacetato	1,2-ethanediyl(dinitrilo)tetraacetato
py	pyridine	pyridine
thf	tetrahydrofuran	tetrahydrofuran
terpy	2,2′,2′-terpyridine	2,2′:6′,2″-terpyridine
dien	diethylenetriamine	*N*-(2-aminoethyl)-1,2-ethanediamine
en	ethylenediamine	1,2-ethanediamine
pn	propylenediamine	1,2-propanediamine
dppe	1,2-bis(diphenyl-phosphino)ethane	1,2-ethanediyl-bis(diphenyl-phosphine)
cod	cyclooctadiene	1,5-cyclooctadiene
18–crown–6	1,4,7,10,13,16-hexaoxacyclooctadecane	1,4,7,10,13,16-hexaoxacyclooctadecane
[12]aneS$_3$	1,4,7,10-tetrathiacyclodecane	1,4,7,10-tetrathiacyclodecane

Dichlorobis(trimethylphosphine)platinum(II) $[PtCl_2(PMe_3)_2]$.
In anionic complexes the metal has an -ate ending:
   Tetrachloroplatinate(II) $[PtCl_4]^{2-}$.
Anionic ligands derived from acids have their -ate names replaced by -ato:
   $[Co(NO_3)_2(NH_3)_4]^+$ -dinitratotetramminecobalt(III).

Non-aqueous solvents

Water is an excellent solvent because of its high dielectric constant (which reduces the force experienced by oppositely charged ions separated by the solvent), its Lewis base capability, and its ability to autoionize. The following non-aqueous solvents also play an important role in inorganic chemistry:

$$NH_3 \quad H_2SO_4 \quad HF \quad BF_3 \quad SO_2$$

	NH_3	SO_2
m.p./°C	−78	−76
b.p./°C	−33	−10
Dielectric const.	22 at −33°C	17 at −16°C
	(cf. H_2O 79)	

Ammonia is particularly useful for solvating electrons, e.g. sodium metal dissolved in liquid ammonia produces solutions whose blue colour results from electrons trapped in cavities formed by ammonia molecules. Its greater co-ordinating power with metal ions can lead to interesting differences in solubility properties compared with water, e.g.

in water: $KCl + AgNO_3 \rightarrow AgCl\downarrow + KNO_3$
in liquid ammonia: $KNO_3 + AgCl \rightarrow KCl\downarrow + AgNO_3$

The use of ammonia leads to the isolation of salts which are not generally accessible in water. For example, the anion NH_2CONH^- is not formed in aqueous solution because OH^- is not a sufficiently strong base:

in water: $H_2NCONH^- + H_2O \rightarrow OH^- + H_2NCONH_2$
in liquid ammonia: $H_2NCONH_2 + NH_2^- \rightarrow H_2NCONH^- + NH_3$

Neutralization reactions analogous to those observed in water are possible in ammonia because of its ability to autoionize:

$$KNH_2 + NH_4I \rightleftharpoons KI + 2NH_3$$
$$cf. \ KOH + HI \rightleftharpoons KI + H_2O$$

$$Zn^{2+} + 2NH_2^- \rightleftharpoons Zn(NH_2)_2 + 2NH_2^- \rightleftharpoons [Zn(NH_2)_4]^{2-}$$

$$cf. \ Zn^{2+} + 2OH^- \rightleftharpoons Zn(OH)_2 + 2OH^- \rightleftharpoons [Zn(OH)_4]^{2-}$$

Non-protic solvents such as liquid SO_2 are particularly useful for dissolving iodide salts, forming $I \rightarrow SO_2^-$ complexes. It can also behave as an inert carrier for oxidation reduction processes because of its facile reduction to SO_2^- e.g.

in liquid SO_2: $2FeCl_3 + 2KI \rightarrow 2FeCl_2 + 2KCl + I_2$

O

Orbitals

Atomic orbitals

An atomic orbital is the wavefunction (Ψ) of an electron in an atom. Its square (Ψ^2) gives the probability of finding the electron at that point.

The Schrödinger solution defines the wave equation for the electron in the hydrogen atom as a product of a radial and angular part as follows:

$$\Psi_{n,l,m}(r,\theta,\phi) = R_{n,l}(r) \cdot Y_{l,m}(\theta,\phi)$$
$$\text{Radial part} \quad \text{Angular part}$$

where the radial part is governed by the quantum numbers n and l and and the angular part by the quantum numbers l and m. In addition if the equation is solved with relativistic corrections for the motion of the electron around the nucleus then a fourth quantum number s emerges which can take values of + or − 1/2. This quantum number is described as the spin quantum number and is pictorially associated with the ability of the electron to spin on its own axis.

The quantum number n is the principal quantum number which determines the energies of the solutions to the hydrogen atom

$$E = -hcR/n^2$$

A node in a wavefunction is a point, plane, or surface where the electron probability is zero because it represents the intersection of regions of the wavefunction with + and − signs.

where R is the Rydberg constant, c is the speed of light and h is Planck's constant. n can take any positive integer values: 1,2,3,......., and is independent of l, m and s. l is the orbital angular momentum quantum number (sometimes called the azimuthal quantum number in the older textbooks) and can take up integer values from 0 to $(n-1)$. The total angular momentum associated with the wave function $\Psi_{n,l,m}(r,\theta,\phi)$ is $\{l(l+1)\}^{1/2}h/2\pi$ and is related to the number of nodes in the angular part of the wave function. Therefore the shape of the orbital is specified by the number of angular nodes. m is the magnetic quantum number and gives components of angular momentum $mh/2\pi$ along the z axis and is restricted to having $2l+1$ integer values, 0, ± 1, ...± l. This quantum number defines the direction of the orbital in space by specifying the number of nodal planes which coincide in their direction with the z axis. For example, p_x, p_y, and p_z each have the 'dumbell' shape illustrated in the diagram in the margin on p. 71 and a single nodal plane, but only p_x and p_y ($m_l = \pm 1$) have a single nodal plane coincident with the z axis, p_z ($m_l = 0$) has its nodal plane in the xy plane. The choice of direction is arbitrary because all the orbitals are equivalent, but if a magnetic field is applied in a specific direction then this equality is lost and the interaction of the orbital angular momentum with the magnetic field depends on the quantum number m_l.

Ionization potential of H atom to $n = \infty$

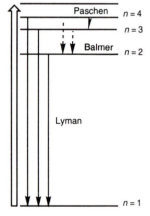

Energy states of the hydrogen atom and the electronic transitions which give rise to the line spectrum of the hydrogen atom. The ionization potential is indicated by the wide arrow on the left.

Therefore, each energy state of the electron is labelled by four quantum numbers. Three quantum numbers n, l , and m define an orbital which can accommodate an electron with a spin of +1/2 or −1/2. All orbitals with the same quantum number n are said to belong to the same *shell* and all orbitals sharing the same quantum number l are said to belong to the same *subshell*. Therefore each shell has a total of n^2 orbitals all with the same energy and each subshell has $2l+1$ orbitals. The subshells are designated the letters s, p, d, f, etc. to describe their common quantum number l as follows:

$l =$	0	1	2	3	4
	s	p	d	f	g
number of orbitals		1	3	5	7	9

n	l	subshell	m_l	No. of orbitals in subshell
1	0	1s	0	1
2	0	2s	0	1
	1	2p	1,0,–1	3
3	0	3s	0	1
	1	3p	1,0,–1	3
	2	3d	2,1,0,–1,–2	5
4	0	4s	0	1
	1	4p	1,0,–1	3
	2	4d	2,1,0,–1,–2	5
	3	4f	3,2,1,0,–1,–2,–3	7

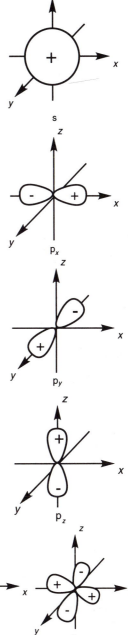

In any one shell there is one s orbital, three p orbitals (if $n >1$), five d orbitals (if $n >2$), and seven f orbitals (if $n >3$). Since the subshell orbitals are associated with different numbers of nodes they have different shapes and these are illustrated in the diagram. It is noteworthy that the s orbital ($l=0$) has no angular nodes and is therefore spherical, the three p orbitals have a single node and may be viewed as dumbells pointing along the $x,y,$ and z directions, the five d orbitals have two nodes each. Since there are not five independent and doubly noded functions the $m = 0$ solution has two nodal cones intersecting at the origin and spreading out along the + and $-z$ directions. The remaining four d functions have two nodal planes intersecting at right angles at the nucleus. The total number of angular nodes is therefore equal to l.

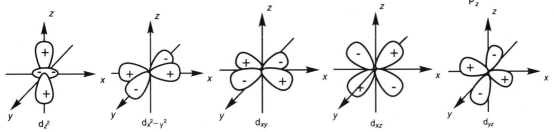

The diagrams on this page are schematic illustrations of the angular parts of the wave functions for the hydrogen atom for $l =$ 0,1, and 2.

Points at *x,y,z* and *-x,-y,-z* are identical and related by a centre of inversion

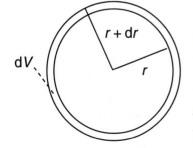

Orbitals which have a centre of inversion are gerade **g** and those which do not are ungerade **u** .

A periodic table based upon the Schrödinger solution to the hydrogen atom: schematic illustration of the energy states of the hydrogen atom. On the right hand side the box diagram indicates the periodic table which would result if this energy level scheme were valid for all atoms. Since the periodic table does not have this structure this means that the energy states for polyelectronic atoms must differ from those of hydrogen (see *Periodic table*).

The shells, subshells, and energies of the solutions of the Schrödinger equation are illustrated in the diagram below.

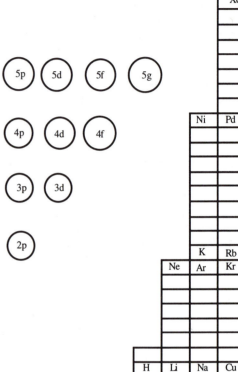

No of elements which can be fitted in row

| | | 2 | 8 | 18 | 32 | 50 |

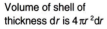

Volume of shell of thickness d*r* is $4\pi r^2 dr$

The radial part of the wavefunction $R_{n,l}(r)$. is related to the probability of finding the electron in the region of the nucleus. More specifically the radial distribution function $4\pi r^2 \Psi^2 dr$ is the probability of finding the electron in the shell of thickness d*r* located at *r* from the nucleus. The surface area of the spherical shell increases with *r* hence the probability is zero at the nucleus passes through more than one maximum and decays exponentially at longer distances. Each time the wavefunction passes through zero along the *r* axis corresponds to a radial node in the wavefunction (except when *r* = 0). The total number of radial nodes for an orbital with the quantum numbers *n* and *l* is *n*–*l*–1. A radial node is therefore a spherical surface separating parts of the radial wavefunction having opposite signs.

The radial distribution functions for the hydrogen atom for *n* = 1, 2, and 3 are shown in the figure opposite. The following points are noteworthy:

The s wavefunctions have $(n-1)$ radial nodes, the p wavefunctions have $(n-2)$ nodes, and the d wavefunctions $(n-3)$ nodes.

For a given l quantum number the maximum in the radial distribution function moves progressively to longer distances from the nucleus. This may be seen most clearly for the 1s, 2s, and 3s orbitals in the diagram below.

For a given n quantum number the maximum in the radial distribution function is farthest from the nucleus for s and comes progressively closer to the nucleus for p and d.

For this subset of orbitals the total electron density within 300 pm of the nucleus is s > p > d. This difference in electron distribution is very important for understanding the different energies of s, p, and d orbitals in polyelectron atoms (see *Periodic table*). For the higher symmetry hydrogen atom the energies of the orbitals are determined solely by the total number of nodes. The total number of radial and angular nodes for a shell is $n-1$ and therefore the subshell orbitals ns, np, nd,... have the same energy (see the large diagram on p. 72).

Interestingly, if the same energy scheme applied to all atoms and the energy levels were filled according to the *Aufbau* procedure the periodic table would have an unfamiliar form. This is illustrated schematically in the same diagram. For the third shell the filling of 3s and 3p would lead to the elements Na–Ar and then filling of the 3d shell would require the elements K, Ca, Sc, Ti, ...Co, Ni to be considered together. The fourth shell would start with Cu corresponding to the filling of the 4s orbital. The hypothetical periodic table also separates Li and Na from K and Rb. It follows that the energy level scheme must change for polyelectronic atoms and this aspect is discussed fully under *Periodic table*.

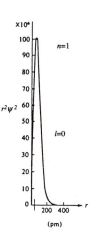

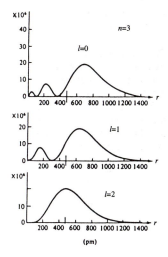

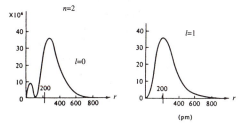

Radial probability functions for $n = 1$, 2 and 3 for the hydrogen atom. The function gives the probability of finding the electron in a spherical shell of thickness dr at a distance r from the nucleus

Organometallic compound

A compound containing a metal carbon bond, which may be a σ-bond, e.g. $Ti(CH_3)_4$ or a π-bond $[PtCl_3(\eta^2\text{-}C_2H_4)]^-$, η^2 indicating that two carbon atoms are bonded to the platinum atom, e.g. $Fe(\eta^5\text{-}C_5H_5)_2$ indicates that the ligand is π-bonded and that all five carbon atoms of both cyclopentadienyl ligands are bonded to the iron atom. Formulae are often abbreviated to $Fe(\eta\text{-}C_5H_5)_2$, $Cr(\eta\text{-}C_6H_6)_2$ if all the carbon atoms are bonded. $Ru(\eta\text{-}C_6H_6)(\eta^4\text{-}C_6H_6)$ has one ring with all carbons π-bonded with the second ring having only four contiguous atoms bonded to the central ruthenium atom.

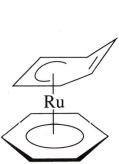

Oxidation state

This is a formal device for partitioning electrons in a molecule in a chemically intelligent way. The oxidation state of an atom in a compound is the charge which would result if the electrons in each bond to that atom were assigned to the more electronegative atom. The more electronegative atom is thereby made to complete its octet of electrons.

Examples:

MnO_4^- Mn^{7+} and $4 O^{2-}$ hence oxidation states Mn^{VII}, O^{-II}

When the central atom has a high electronegativity the formal oxidation state may change dramatically as a function of ligand:

NH_4^+ N^{3-} and $4H^+$ but NF_4^+ N^{5+} and $4F^-$

By convention hydrogen is considered positive in combination with non-metals and negative in combination with metals. This convention can at times lead to a false impression that protonation leads to oxidation of the metal ion.

Example:

$$Co(CO)_4^- \quad + \quad H^+ \quad \rightarrow \quad [Co(CO)_4H]$$
$$(Co^{-I}) \qquad\qquad\qquad\qquad\qquad Co^{I}$$

Ligands which can have an independent existence as molecules are removed with their donor electron pairs, as in the following examples:

$[PtCl_2(NH_3)_2]$	Pt^{II} $2Cl^{-I}$	$2NH_3$
$[Ni(CO)_4]$	Ni^0 $4CO$	
$[PtCl_3(\eta-C_2H_4)]^-$	Pt^{II} $3Cl^{-I}$	C_2H_4

π-bonded cyclically delocalized ligands are removed as their aromatic (4π +2 electrons) molecule or ion; e.g. $C_5H_5^-$, C_6H_6, $C_7H_7^+$.

For a molecule with a homopolar bond the electrons are assigned equally to the atoms. It follows that an element–element bond makes no contribution to the oxidation state.

Examples:

P_4		P^0
P_2H_4	P^{-II}	H^I
$Mn_2(CO)_{10}$	Mn^0	CO

When elements form a range of compounds containing element–element bonds it is therefore more helpful to stress the valency of the atom rather than the formal oxidation state of the ion.

For example, silicon in $SiCl_4$, Si_2R_6, and Si_2R_4 has formal oxidation states of (IV), (III), and (II) but maintains a valency of 4.

Difficulties in assigning oxidation states arise if the elements have similar electronegativities, e.g. NCl_3, N_4S_4; odd electron ligands, e.g. NO and 'non-innocent' ligands.

Non-innocent ligands can accept electrons into low-lying antibonding orbitals, e.g. bipyridyl which may also exist as bipy$^-$. In $[Cr(bipy)_3]^{3+}$ the oxidation state is Cr^{III}, but in $[Cr(bipy)_3]$ it may be either Cr^0 + 3bipy or Cr^{III} + 3bipy$^-$ and only further spectroscopic data can resolve the issue. The ligand $S_2C_2(CN)_2$ presents a similar ambiguity.

More examples

$[Ni(CN)_4]^{2-}$: Ni^{II} $4CN^-$

K_2CrCl_4: $2K^I$ Cr^{II} $4Cl^{-I}$

$BaTiO_3$: Ba^{II} Ti^{IV} $3O^{-II}$

Mn_3O_4: Mn^{II} Mn^{III} $4O^{-II}$

$LaCoO_3$: La^{III} Co^{III} $3O^{-II}$

$CsAu$: Cs^I Au^{-I}

Periodic table (see inside front cover)

Although historically the periodic table can be traced back to Mendeleev in 1871, its modern interpretation depends on the quantum mechanical solution of the wavefunctions of the hydrogen atom and its extension to polyelectronic atoms. Initially Mendeleev's classification depended on similarities in the formulae of chemical compounds, e.g. Li, Na, K all formed simple compounds of formula MX (where X = F, Cl, Br, and I) and the properties of these compounds seemed to vary in a systematic way as the atomic number of the metal atom increased. In addition the metals themselves are all soft, malleable, and highly reactive. Similar trends in properties were established for other families of elements. The arrangement of elements in order of increasing atomic number with the elements showing similarities placed in vertical columns formed the basis of the original periodic table. We now realize that the fundamental basis of this periodicity is the repeating character of electron configurations of atoms resulting from the *Aufbau* filling of quantized energy levels for polyelectronic atoms.

In *Orbitals* the solution of the wavefunctions of the hydrogen atom according to the Schrödinger equation was discussed in some detail and it was shown that for this specific case the energies of the orbitals ns, np, nd, etc., had the same energies and in this particularly simple case they depend solely on the principal quantum number n.

Polyelectronic atoms

In polyelectronic atoms it is assumed that the angular parts of the wave functions are identical to those of the hydrogen atom described in *Orbitals*, but the radial part is altered to take into account the differences in nuclear charge and electron repulsion experienced by the electrons. Indeed the orbitals within a subshell differ in energy in a polyelectronic atom because of differential screening and penetration effects which their electrons experience.

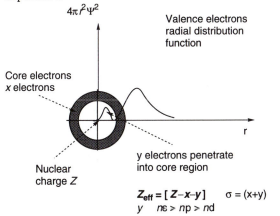

$$Z_{eff} = [Z-x-y] \quad \sigma = (x+y)$$
$$y \quad ns > np > nd$$

Screening and penetration

The outer electrons in an atom do not experience the full weight of the nuclear charge associated with the nucleus because of the shells and subshells

P

Effective nuclear charges (Z_{eff}) of valence orbitals (Clementi–Raimondi values).

	ns	np
Li	1.28	
Na	2.51	
K	3.50	
B	2.58	2.42
Al	4.12	4.07
Ga	7.07	6.22
C	3.22	3.14
Si	4.90	4.29
Ge	8.04	6.78
N	3.85	3.83
O	4.49	4.45
F	5.13	5.10

Note that the effective nuclear charge for $ns > np$ and the difference increases as n becomes larger. Also the effective nuclear charge increases across the series Li–F because as the additional electrons are introduced they do not screen the valence electrons from the nucleus with perfect efficiency.

The radial distribution function of the valence electrons shown schematically on the left has a part which lies on the outside of the spherical shell representing the core electrons and a part which penetrates. That part which remains outside the core experiences an effective nuclear charge:
$Z_{eff} = Z-x-y$, where Z is the nuclear charge, y is the number of electrons which make up the core, and x is the proportion of electron density associated with the radial distribution function which penetrates the core.

of electrons lying between them and the nucleus 'screen' them. Therefore, the outer valence electrons experience an effective nuclear charge, $Z_{eff} = Z - \sigma$, where σ is the screening constant. This screening constant depends on the degree of penetration of the orbital inside the radial distribution functions of the core electrons. The differences in the radial distribution functions for ns, np, nd, and nf orbitals arising from the $n-l-1$ radial nodes (see the Figure on p. 73) means that their relative penetrating power is s>p>d>f and therefore the effective nuclear charge for $ns>np>nd>nf$. (See table in margin on p. 78). Consequently in polyelectronic atoms the subshells no longer are of equal energy and their relative energies are $ns>np>nd>nf$. This change in the relative ordering of the atomic orbitals has a very profound influence on the structure of the periodic table. The differences in screening and penetration leads to the following energy level ordering for most atoms.

In this Figure the energy levels of the atoms are indicated on a vertical scale and filled in the sequence indicated by the arrows. The orbital fillings associated with this Aufbau procedure is indicated in the box diagram on the right. Each box represents an orbital which can accommodate two electrons. The orbitals being filled are labelled accordingly. After 6s the 4f orbitals are filled and since we have run out of dimensions for displaying the different shells this is represented by the horizontal dotted line and similarly after 7s the 5f orbitals are filled. When these are completely full then the 5d and 6d orbitals are occupied respectively. This box diagram is therefore a schematic representation of the conventional periodic table shown on the inside front cover when it is rotated clockwize by 90°, two elements occupying the space of each box.

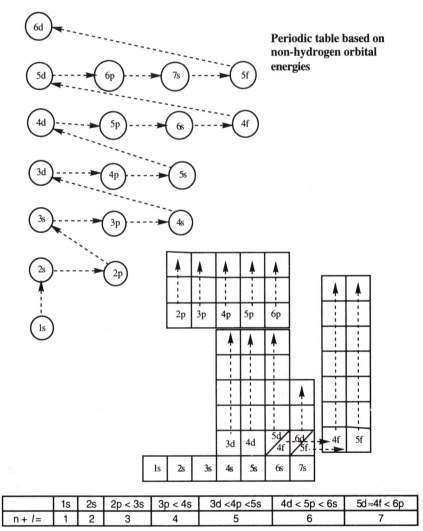

Periodic table based on non-hydrogen orbital energies

	1s	2s	2p < 3s	3p < 4s	3d <4p <5s	4d < 5p < 6s	5d ≈4f < 6p
$n + l =$	1	2	3	4	5	6	7

For the same $n + l$, $d > f > p > s$ usually

The universal energy level diagram shown in the Figure can account for most of the features of the periodic table as we currently know it. The energy level diagram shown in this Figure is again filled according to the *Aufbau principle*, i.e. orbitals with the lowest energy are filled first. The Pauli exclusion principle which states that no two electrons may have the same four quantum numbers also operates to limit the total number of electrons to subshells as shown in the table opposite.

The relationship between this energy level ordering diagram and the periodic table is shown in the Figure above and the electron configurations of the atoms in their ground states are given in the table on the inside back cover. The electronic description of the periodic table readily accounts for its identifiable regions in terms of common outer configurations of atoms. The relevant periodic groups are summarized in the margin.

The *Aufbau principle* developed above is based on the premise that successive orbitals have quite large energy separations and electron repulsion effects are unimportant. When these conditions are not met electron configurations which diverge from those predicted strictly according to the *Aufbau principle* are observed.

Those in italics in the margin on p. 78 represent exceptions which can be attributed to exchange energy effects and are associated with either full or half full d and f shells (see *Exchange energy*). The remainder are associated with the similar orbital energies of 4d and 5s, 5d and 6s, 5d and 4f, and 6d and 5f orbitals, and the effects of interelectronic repulsion. Repulsion between two electrons in different orbitals is smaller than that between two electrons in the same orbital. If the energy difference between two orbitals is smaller than the difference in electron repulsion for the two electron configurations then there is energy gain by the two electrons occupying different orbitals.

A modern periodic chart of the elements is shown on the inside front cover and includes the IUPAC recommended numbering of the groups. The s block elements (alkali metals and alkaline earths) make up groups 1 and 2. The d block elements (the transition metals) make up groups 3–12 and are associated with the filling of the d shells. The p block elements (the post-transition elements of which the non-metals are a sub-group) represent columns 13–18. The f block elements are not numbered in columns within the recommended IUPAC classification primarily because the lanthanides and actinides have relatively similar properties. The older 'A' and 'B' classification of the periodic groups is also illustrated in this table.

The chemical implications of the periodic table arise primarily from the fact that elements with the same numbers of valence electrons occupying orbitals with the same l quantum numbers form compounds with identical stoichiometries.

The radial part of the wavefunction is very important in deciding whether the orbitals overlap with orbitals of other atoms. It is difficult to make sweeping generalizations but the following may be helpful. The atomic orbitals of the first short period of elements (2s and 2p) overlap particularly effectively with the orbitals of adjacent atoms leading to strong σ- and π-bonds.

subshell	total number of electrons
s	2
p	6
d	10
f	14

Group 1 Alkali metals

Li–Fr $\quad ns^1 \quad (n = 2\text{–}7)$

Group 2 Alkaline earth metals

Be–Ra $\quad ns^2 \quad (n = 2\text{–}7)$

Post-transition elements

B–Ne $\quad 2s^2 2p^1 \text{–} 2s^2 2p^6$

Al–Ar $\quad 3s^2 3p^1 \text{–} 3s^2 3p^6$

Ga–Kr $\quad 4s^2 4p^1 \text{–} 4s^2 4p^6$

In–Xe $\quad 5s^2 5p^1 \text{–} 5s^2 5p^6$

Tl–Rn $\quad 6s^2 6p^1 \text{–} 6s^2 6p^6$

Group 15 Pnictonides $ns^2 np^3$
Group 16 Chalcogenides $ns^2 np^4$
Group 17 Halogens $ns^2 np^5$
Group 18 Inert (noble) gases $ns^2 np^6$

First transition series

Sc–Zn $\quad 4s^2 3d^1 \text{–} 4s^2 3d^{10}$

Second transition series

Y–Cd $\quad 5s^2 4d^1 \text{–} 5s^2 4d^{10}$

Third transition series

La–Hg $\quad 6s^2 5d^1 \text{–} 6s^2 5d^{10}$

Lanthanides Ce–Lu $4f^1 \text{–} 4f^{14}$

Actinides Th–Lr $5f^1 \text{–} 5f^{14}$

Exceptions to regular fillings of orbitals are shown on page 78.

Examples of exceptions
to the Aufbau principle

$Cr\ 3d^5 4s^1$ $La\ 5d^1 6s^2$
$Cu\ 3d^{10} 4s^1$ $Ce\ 4f^1 5d^1 6s^2$
$Gd\ 4f^7 5d 6s^2$ $Pr\ 4f^3 6s^2$
$Mo\ 4d^5 5s^1$ $Nb\ 4d^4 5s^1$

$Pd\ 4d^{10} 5s^0$ $Ru\ 4d^7 5s^1$
$Ag\ 4d^{10} 5s^1$ $Rh\ 4d^8 5s^1$
$Au\ 5d^{10} 6s^1$ $Th\ 6d^2 7s^2$
$Cm\ 5f^7 6d^1 7s^2$ $Pa\ 5f^2 6d^1 7s^2$
$Lr\ 5f^{14} 6d^1 7s^2$ $Np\ 5f^4 6d^1 7s^2$
                 $Pt\ 5d^9 6s^1$

Effective nuclear charges for the valence orbitals of the transition and post-transition elements.

	4s	3d	4p
Ti	4.82	8.14	
Mn	5.28	10.53	
Ni	5.71	12.53	
Zn	5.97	13.88	
As	8.94	17.38	7.45

Note the much larger effective nuclear charges experienced by the d orbitals which results in their contraction. Note also the steady increase in Z_{eff} across the transition series because of the inefficient screening abilities of the incoming electrons.

For the heavier post-transition elements the overlap becomes less effective because of the increased size of the atoms and the presence of inner shells. The π-bonding is particularly affected by this decrease in overlap ability. The 3d orbitals of the transition metals are contracted relative to 4s and 4p (see table in margin for examples of effective nuclear charges) and therefore only form strong covalent bonds when the distances to the ligands are short. The overlap ability of the 4d and 5d orbitals is greater than that of 3d leading to more covalency. The 4f orbitals are very contracted relative to the other valence orbitals and therefore are almost core-like and do not contribute greatly to covalent bond formation. The 5f orbitals have a more favourable radial extension and therefore the actinide compounds show more covalency than those of the lanthanides.

Periodic trends

Vertical trends

Down a column of the periodic table the atoms share a common electronic configuration and therefore tend to form compounds with the same stoichiometry, e.g. B_2O_3, Al_2O_3, Ga_2O_3, In_2O_3, and Tl_2O_3. However, the changes in the sizes of the atoms, the radial extent of the orbitals, and ionization energies can lead to these series of compounds having very different properties. For example, both CO_2 and SiO_2 share the same stoichiometry but the former has a molecular structure with C–O multiple bonds and the latter has an infinite structure based on tetrahedral SiO_4 fragments. Mindful of these anomalies one can still discern the gradual evolution of properties down any group of the periodic table.

For the majority of elements several oxidation states are observed and their relative stabilities change as one descends the column of the periodic table. For example aluminium is only observed in compounds with oxidation state (+3), but for the heavier elements Ga, In, Tl the lower oxidation state (+1) becomes progressively more stable and more compounds with this oxidation state are observed (see *Inert pair effect*). This pattern of behaviour is also observed for Si–Pb and P–Bi, but for the halogens and noble gases the trend is reversed and the higher oxidation state becomes more stable. Indeed for the noble gases stable compounds are restricted to xenon.

In the transition metal series the higher oxidation states become more accessible as one descends the column of the periodic table. For example, iron is observed as (+6) in $[FeO_4]^{2-}$ but not in the (+8) oxidation state, in contrast ruthenium and osmium form RuO_4 and OsO_4.

These trends may be rationalized within the framework of the modern quantum-mechanically based periodic table in terms of differences in ionization potentials, promotion energies, and relative bond enthalpies.

Horizontal trends

Molecular compounds formed by adjacent elements in the periodic table can be closely related by isoelectronic analogies. For example, the noble gas compounds would have been discovered many years earlier if more account had been taken of the isoelectronic relationships between

$[IF_2]^-$ and XeF_2

$[IF_4]^-$ and XeF_4 $[IO_4]^-$ and XeO_4.

In the transition series metal ions with common electronic configurations form similar co-ordination compounds, e.g. Au^{III}, Pt^{II} and Rh^{I} all form square-planar d^8 16 electron complexes.

When the valence orbitals are effectively core-like for a sub-shell of elements then one obtains a series of compounds which have the same formula and similar properties. This effect is particularly noticeable for the first transition series in their compounds with electronegative ligands, the lanthanides and the actinides.

All the lanthanides form M_2O_3 and MCl_3 compounds which have similar structures and the M^{3+} metal ions differ only by having progressively filled 4f orbitals which are so contracted in size that they participate little in bonding. The actinides show similar series of compounds although the 5f orbitals are less core like than the 4f orbitals.

Diagonal trends

Although the *periodic table* emphasizes chemical relationships in columns there also exist some diagonal similarities in chemical properties. For example, the pairs of elements, Li and Mg, Be and Al, and B and Si share many similar chemical properties. These similarities may be interpreted in terms of their similar *polarizabilities*.

Lithium–magnesium

Both react directly with N_2 to form nitrides; Li_3N, Mg_3N_2. The solubilities of lithium salts resemble those of magnesium. The organometallics LiR and MgRX are both used in organic chemistry as convenient sources of R^-.

Li	
	Mg

Beryllium–aluminium

Both form very stable insoluble oxides, thereby rendering them passive to acids. They have similar standard reduction potentials, $E^{\ominus}_{Be^{2+}/Be} = -1.85$ V, $E^{\ominus}_{Al^{3+}/Al} = -1.66$ V. Their carbonates have similar stabilities and their oxides are amphoteric.

Be	
	Al

Boron–silicon

Both elements are non-metallic, have volatile and very reactive hydrides, the halides (except BF_3) readily hydrolyse to form boric or silicic acid. Both form a wide range of borate/silicate salts based upon oligomers of the oxo-anions.

B	
	Si

Ru	Rh	Pd
Os	Ir	Pt

Platinum metals

The heavier elements of groups 6, 7, and 8; Ru, Rh, Pd, Os, Ir, and Pt.

Polarizability

When an electric field is applied to an atom or molecule the electron distribution is modified and they acquire an electric dipole moment. The moment (μ) is given by:

$$\mu = \alpha E$$

where α is the polarizability of the molecule, and E is the electric field strength. The electronic polarizability of an atom increases with atomic radius and with the number of electrons it contains. It is easier to distort the electronic distribution when the electrons are far from the nucleus or well shielded from its charge.

In inorganic chemistry the polarizability of ions is related to the degree of covalency through *Fajans' rules*.

Fajans' rules

1. High charge and small size of cation
This combination exerts a strong electric field on an anion and polarizes it towards the cation, thereby increasing the degree of covalency. The ratio of the cation's oxidation state and its ionic radius (Z/r) is described as the ionic potential. Some typical values are given opposite.

Li^+, 17 Be^{2+}, 64 B^{3+}, 156
Na^+, 10 Mg^{2+}, 31 Al^{3+}, 60
K^+, 8 Ca^{2+}, 20 Ga^{3+}, 48

The effects of covalency are obvious in the following m.ps:

$BeCl_2$ $CaCl_2$
405°C   772°C

$NaBr$ $MgBr_2$ $AlBr_3$
775°C   700°C   97.5°C

LiF $LiCl$ $LiBr$ LiI
870°C   613°C   547°C   446°C

$CaCl_2$ $HgCl_2$
772°C   276°C

These data may be used to rationalize the diagonal relationships.

2. High charge and large size of anion
The polarizability of an anion is related to its size and charge and therefore the following are particularly polarizable:

$$I^-\quad Se^{2-}\quad Te^{2-}\quad As^{3-}\quad P^{3-}$$

Compounds formed from these ions will therefore have a high degree of covalency.

3. Electron configuration of cation
Two cations may have similar radii and charges (e.g. Ca^{2+} 114 pm and Hg^{2+} 116 pm) but different effective nuclear charges because their outer subshells have different shielding abilities. In particular an ion with $(n-1)\,d^x\,ns^0np^0$ has a higher effective nuclear charge than an ion with $(n-1)s^2\,(n-1)p^6\,ns^0$ and consequently is more polarizing. The group 11 and 12 metals therefore form compounds which are more covalent, are less soluble in water, and form more stable co-ordination compounds than the group 1 and 2 metals of the same size and charge.

Polyhedral cage geometries

Many elements form molecules with element–element bonds which are described as cage, cluster, or polyhedral molecules. The geometries of these molecules are described in terms of the following idealized classes of polyhedra.

Deltahedra

The deltahedra have exclusively triangular faces and are illustrated below for $n = 4$–12. These polyhedra are adopted by electron deficient clusters, e.g. $B_6H_6^{2-}$ and metal carbonyl clusters, e.g. $[Co_6(CO)_{15}]^{2-}$. These deltahedral geometries are described as *closo*-structures, emphasizing the closed (complete nature) of the polyhedron.

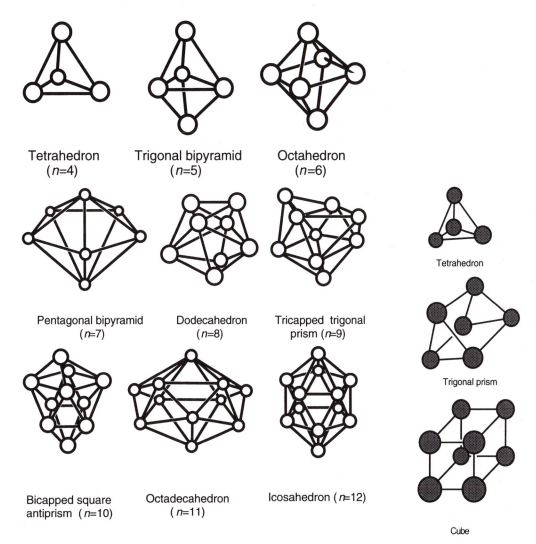

Tetrahedron (*n*=4)

Trigonal bipyramid (*n*=5)

Octahedron (*n*=6)

Pentagonal bipyramid (*n*=7)

Dodecahedron (*n*=8)

Tricapped trigonal prism (*n*=9)

Bicapped square antiprism (*n*=10)

Octadecahedron (*n*=11)

Icosahedron (*n*=12)

Tetrahedron

Trigonal prism

Cube

Three-connected polyhedra

These polyhedra have three connectivities at each of the vertices and therefore can only exist when there are an even number of vertices. Typical examples are illustrated in the margin. These cage compounds are preferred by atoms and molecular fragments which favour the formation of three localized two-centre two-electron bonds, e.g. CH, P, or $Co(CO)_3$.

Those high-symmetry polyhedra which have all the faces identical and also the vertices identical are described as Platonic polyhedra, i.e tetrahedron, cube, octahedron, icosahedron, and dodecahedron.

Cuneane

Examples of 3-connected polyhedra

R

	H		
	37		
C	N	O	F
77	74	74	72
Si	P	S	Cl
118	110	104	100
Ge	As	Se	Br
122	121	117	114
	Sb	Te	I
	141	137	133

N.B. There is a particularly large change in radii in going from the second period to the third. This has great implications for the co-ordination numbers and reactivities of these elements.

N.B. When several oxidation states are observed for an atom it is necessary to define an ionic radius for each. The corresponding contraction can be enormous, e.g. I^- (220 pm) to I^{VII} (53 pm).

Radii

Covalent radii

The covalent radius is half the distance between like atoms in a molecule where the bond order is well defined. For example, the single, double, and triple bond covalent radii of carbon may be derived from the observed C–C bond lengths in C_2H_6, C_2H_4, and C_2H_2.

Covalent radii are sufficiently additive that heteronuclear bond lengths may be estimated with some accuracy from the covalent radii of the atoms which are joined, e.g. PI_3 has a P–I bond length of 243 pm which is close to the sum of the covalent radii of P (110 pm) and I (133 pm).

Deviations from this simple additive relationship occur when:

(a) the co-ordination number changes: Cl–O in ClO_2^- (156 pm), ClO_3^- (149 pm), and ClO_4^- (143 pm);

(b) the hybridization changes: C–H sp^3 74, sp^2 73, sp 69 pm;

(c) ionic character of the bond, e.g. the Si–O distance in SiO_2 (~160 pm) is much shorter than the sum of the covalent radii (191 pm);

(d) some degree of multiple bond character, e.g. the O–O bond length in O_2F_2 (122 pm) is much shorter than that in O_2H_2 (146 pm) because of a contribution from the *resonance* forms:

$$F^- \quad O=O^+–F \qquad \leftrightarrow \qquad F–O^+=O \quad F^-$$

Ionic radii

The distance between cation and anion in an infinite lattice may be defined experimentally with great accuracy. The contribution of the anion and the cation to this total requires theoretical interpretation and a wide range of possible models have been proposed, e.g. the cation and the anion contributions are related inversely to the effective nuclear charges of the constituent atoms.

Ionic radii for the main group elements. Numbers in italics are oxidation states. All values are for co-ordination number 6 (except 4 for N^{3-})

H	Li	Be	B	C	N	O	F
−1 ~150	*+1* 76	*+2* 45	*+2* 27	*+4* 16	*−3* 146	*−2* 140	*−1* 133
					+3 16		
	Na	Mg	Al	Si	P	S	Cl
	+1 102	*+2* 72	*+3* 54	*+4* 40	*+3* 44	*−2* 184	*−1* 181
					+5 38	*+6* 29	
	K	Ca	Ga	Ge	As	Se	Br
	+1 138	*+2* 100	*+3* 62	*+2* 73	*+3* 58	*−2* 198	*−1* 196
				+4 53	*+5* 46	*+4* 50	
	Rb	Sr	In	Sn	Sb	Te	I
	+1 152	*+2* 118	*+3* 80	*+2* 118	*+3* 76	*−2* 221	*−1* 220
				+4 69	*+5* 60	*+4* 97	*+5* 95
						+6 56	*+7* 53
	Cs	Ba	Tl	Pb	Bi	Po	
	+1 167	*+2* 135	*+1* 150	*+2* 119	*+3* 103	*+4* 94	
			+3 89	*+4* 78	*+5* 76	*+6* 67	

If the radii are genuinely additive then once one radius has been defined then the radii of all atoms can be defined with respect to it. The commonly used values for ionic radii are based on the radius of O^{2-} as 140 pm. These data which were first critically valued by Shannon are summarized in the tables opposite (bottom of page) and below.

For transition metals the radius depends on whether the ion has a low (ls) or high spin (hs) configuration (see *Ligand field theory*). The radius is smaller for low spin because the antibonding e_g orbitals which point directly at the ligands in an octahedron are vacant.

Ionic radii for transition elements. Numbers with signs: oxidation states; ls = low spin, hs = high spin; co-ordination numbers in roman numerals if other than 6.

	Sc	Ti	V	Cr	Mn	Fe	Co	Ni	Cu	Zn	
+2				ls 73	ls 67	ls 61	ls 65		+1 77		
+2		86	79	hs 80	hs 83	hs 78	hs 75	69	73	74	+2
+3	75	67	64	62	ls 58	ls 55	ls 55	ls 56	ls 54		+3
+3					hs 65	hs 65	hs 61	hs 60			+3
+4		61	58	55	53	59	hs 53	ls 48			+4
+5		543	49	IV 26							+5
+6			44	IV 25	IV 25						+6

The bond length in a polar covalent molecule such as $SiCl_4$ may be estimated from ionic or covalent radii.

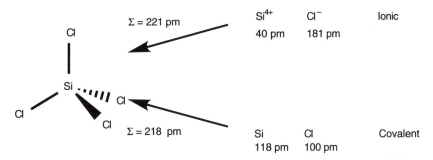

	Si^{4+}	Cl^-	Ionic
$\Sigma = 221$ pm	40 pm	181 pm	

	Si	Cl	Covalent
$\Sigma = 218$ pm	118 pm	100 pm	

In fact both methods overestimate the bond length by 18 pm (±1). The following correction has been proposed to account for the contribution resulting from the ionic character of the bond:

$$r_{AB} = r_A + r_B - 9(\chi_A - \chi_B)$$

where χ_A and χ_B are the electronegativity coefficients of A and B respectively.

As with covalent and metallic radii corrections may be made for situations where the co-ordination number is not 6; the metal is in a different oxidation state, etc.

Metallic radii

This is defined as half the distance between next nearest atoms in a metal in the solid phase. Since metals have 12 nearest neighbours in hcp and ccp structures and 8 in bcc structures, the radii for the latter are corrected to a value which would be expected for 12 coordination. For example sodium which has a bcc structure has an experimentally determined radius of 185 pm but this is corrected to 191 pm for 12 co-ordination. The table below summarizes the metallic radii.

Li	Be																Al	
152	112																	
Na	Mg																Al	
186	160																143	
K	Ca	Sc	Ti	V	Cr	Mn	Fe	Co	Ni	Cu	Zn	Ga						
230	197	162	146	134	128	137	126	125	125	128	134	135						
Rb	Sr	Y	Zr	Nb	Mo	Tc	Ru	Rh	Pd	Ag	Cd	In	Sn					
247	215	180	160	146	139	135	134	134	137	144	151	167	154					
Cs	Ba	La	Hf	Ta	W	Re	Os	Ir	Pt	Au	Hg	Tl	Pb					
267	222	187	158	146	139	137	135	136	139	144	511	171	175					

Ce	Pr	Nd	Pm	Sm	Eu	Gd	Tb	Dy	Ho	Er	Tm	Yb	Lu
182	182	182	181	180	204	179	178	177	176	175	174	193	174
Th	Pa	U	Np	Pu	Am	Cm	Bk	Cf	Es	Fm	Md	No	Lr
180	161	156	155	159	173	174	170	186	186				

Van der Waals radii

In a crystalline compound the molecules are closely packed and the intermolecular distances between adjacent atoms represent a compromize between the attractive and repulsive interatomic forces. The shortest commonly observed intermolecular distance between atoms of the same kind is defined as twice the van der Waals radius of that atom.

Some typical van der Waals radii in pm are summarized in the Table below:

					He 140
H 120	C 170	N 155	O 152	F 147	Ne 154
	Si 210	P 180	S 180	Cl 175	Ar 188
	Ge	As 185	Se 190	Br 185	Kr 202
	Sn	Sb 200	Te 206	I 198	Xe 216

Distances shorter than the van der Waals radii in a crystal indicate the occurrence of unusual attractive forces, e.g. hydrogen bonding or weak covalent bond formation.

Radius ratios

see (*Crystal structures, closest packing of spheres.*)

For ions which are really hard incompressible spheres it is possible to calculate the radius ratios of cation (r_M) and anion (r_X) which achieve optimal packing. Optimal packing means that the cation at the centre of the co-ordination polyhedron touches all the anions surrounding it and all the anions which make up the co-ordination polyhedron are also touching. The means of calculating these radius ratios for tetrahedral, octahedral, and cubic polyhedra are shown on p. 85 and the results for a wider range of polyhedra are summarized in the margin.

Geometric radius ratio calculations

The calculations for 8, 6, and 4 co-ordination may be obtained by defining the co-ordination polyhedron in a cube and calculating $2r_X$ (the distance

Co-ordination geometry	r_M/r_X
Trigonal planar (3)	0.155
Tetrahedral (4)	0.225
Octahedral (6)	0.414
Trigonal prismatic (6)	0.528
Cubic (8)	0.732
Dodecahedral (8)	0.668
Square antiprismatic (8)	0.645
Cuboctahedral (12)	1.000
Anticuboctahedral (12)	1.000

between touching anions) and $r_X + r_M$ (the distance between the cation and anion centres) in terms of the length of the cube edge, a.

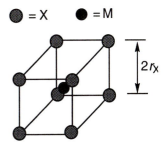

$$= X \qquad \bullet = M$$

8-co-ordination – cube:

$2r_X = a$ (anions in contact along edge of cube)

$r_X + r_M = a\sqrt{3}/2$ (1/2 diagonal of cube)
$\qquad = 2r_X\sqrt{3}/2$

$r_M = \sqrt{3}r_X - r_X = 0.732\ r_X$

M occupies the centre of a cube of side a equal to $2r_X$

6-co-ordination –octahedron:

$2r_X = a$ (anions in contact along face of cube parallel to edge)

$r_X + r_M = a\sqrt{2}/2$ (1/2 diagonal of cube face)
$\qquad = \sqrt{2}/2(2r_X)$

$r_M = \sqrt{2}r_X - r_X = 0.414\ r_X$

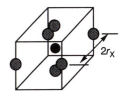

M occupies an octahedral hole in a cube of side a equal to $2\ r_X$

4-co-ordination – tetrahedron:

$2r_X = a\sqrt{2}$ (anions in contact across face of cube)

$r_X + r_M = a\sqrt{3}/2$ (1/2 diagonal of cube)
$\qquad = \sqrt{3}/2(2r_X/\sqrt{2})$

$r_M = \sqrt{3}r_X/\sqrt{2} - r_X = 0.225\ r_X$

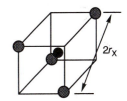

M occupies a tetrahedral hole in a cube of side a, $\sqrt{2}.\,a = 2\,r_X$.

    Since the close contact between the cation and the anion is the more energetically important interaction the radius ratio represents the preferred lower limit, e.g. tetrahedral co-ordination is preferred relative to octahedral for radius ratios which span the range 0.225 to 0.414.

    In real chemical compounds the ions are not incompressible hard spheres, nor are the compounds purely ionic and therefore the radius ratios do not accurately predict the structures of the majority of ionic compounds. Examples of correct predictions include:

Compound	Calculated r_M/r_X	Observed co-ordination geometry
CsCl	0.922	8:8
NaCl	0.563	6:6
ZnS	0.402	4:4

Exceptions include:

Compound	Calculated r_M/r_X	Observed co-ord. geometry	Calculated co-ord. geometry
RbBr	0.84	6:6	8:8
RbCl	0.69	6:6	8:8

    As the covalency in the bond increases bond directionality effects outweigh purely geometric considerations and the radius ratio rules show an increasing number of exceptions.

Resonance

In *valence bond theory* resonance denotes the superposition of wavefunctions to produce a better approximation to the actual wavefunction of the molecule. For example, a starting wavefunction for the HF molecule would have the electrons paired between H and F in a covalent bond and this wavefunction is represented in chemical terms by H–F.

Ionic–covalent resonance describes the polarity of the bond by introducing contributions from the ionic wavefunction represented chemically by $H^+ F^-$. The degree of mixing between the two wavefunctions is computed using the variational theorem but is represented qualitatively by the following symbolism:

$$H\text{–}F \ \leftrightarrow \ H^+ F^-$$

and the mixing is described as *resonance* between these *canonical forms* which represent the relevant wavefunctions. The molecule does not oscillate between these two forms as a function of time. The mixed wavefunctions therefore represent a more accurate representation of the wave function of the actual molecule than either of the wavefunctions of the canonical forms.

Multiple-bond resonance

Describes the degree of multiple bond character in a bond, for example:

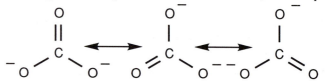

None of the resonance forms of the carbonate structure shown above has an independent existence. The carbonate ion is a symmetrical structure with a three-fold axis of symmetry and each bond has a formal *bond order* of $1^1/_3$. Therefore, it is expected to be shorter than a C–O single bond, but longer than C=O.

These arguments based on resonance delocalization may be used to rationalize the acid strengths of oxo-acids (Pauling's rules) in much the same way as that developed in organic chemistry. Specifically an acid which on dissociation forms an anion where the charge may be delocalized effectively by resonance is stronger than one where this possibility is more limited. For example:

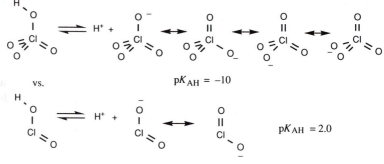

$Li\text{–}H \leftrightarrow Li^+ H^-$

$H_2P\text{–}H \leftrightarrow H_2P^+ \ H^-$

In the former example the large *electronegativity* difference between Li and H (1.22) leads to a significant contribution from the ionic resonance form. In the latter the P and H *electronegativity* difference of 0.02 suggests a minimal contribution from the ionic form.

In general the relative contributions of the resonance forms are estimated from considerations based either on *electronegativity* differences or the *electroneutrality principle*. For example, the contribution from O=C=O is much greater than that from $^+O\equiv C\text{–}O^-$ because the former conforms better to the *electroneutrality* principle. It is noteworthy that the structures chosen for the resonance forms invariably conform to the *inert gas rule*.

Sandwich compounds

Sandwich compounds have two planar cyclic ligands bonded through their faces to a metal atom. Some typical examples are illustrated in the margin:

For transition metal sandwich compounds the *effective atomic number rule* is commonly obeyed. Each C–H fragment donates one π-electron for bonding and the metal contributes all its valence electrons for bonding. Examples, of this procedure are illustrated below.

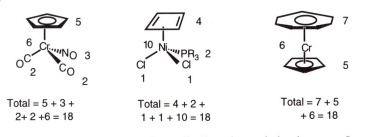

| Total = 5 + 3 + | Total = 4 + 2 + | Total = 7 + 5 |
| 2+ 2 +6 = 18 | 1 + 1 + 10 = 18 | + 6 = 18 |

In these complexes the organic cyclic ligands are behaving as π-Lewis donors to the metal ions.The lanthanides and actinides form sandwich compounds with C_8H_8 but these do not conform to the effective atomic number rule. There are also examples of triple decker sandwiches, half sandwiches, and bent sandwiches as illustrated in the margin.

Stability

In everyday language a stable compound is one which may be stored unchanged in a container over a long period of time. In chemical terms a much more careful definition has to be adopted and it has to be specified whether the compound maintains its integrity for thermodynamic or kinetic reasons. The temperature dependence of the rates of reaction and the *free energies* of reactions also require that the conditions of stability must be defined precisely. For a specific compound MX there are a number of possible decomposition reactions which could lead to its decomposition without the intervention of any external reagents. Such reactions may be described as *intrinsic reactions* and they may have either positive or negative free energies of reaction associated with them. Alternatively, there are reactions involving external reagents and particularly those involving atmospheric gases and H_2O which cause the decomposition of the compound. (Although N_2 is normally regarded as an inert gas there are compounds which are sufficiently reactive to react with it to form either dinitrogen compounds or nitrides.) These may be described as *extrinsic reactions* and may also have positive or negative free energies, but differ from intrinsic reactions because if the compound is isolated from the reactive gases or moisture then it will not decompose. For the intrinsic and extrinsic reactions the rates of the possible decomposition reactions, i.e. the activation energies for the reactions, are also important in deciding whether a compound has a long half-life. Kinetic effects are particularly prone to steric effects of ligands which hinder the attack of the O_2 and H_2O molecules on the central atom. The range of possibilities arising from the interplay of thermodynamic and kinetic factors is summarized in the following table.

S

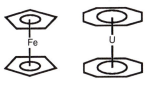

Ferrocene   Uranocene

Cr: $3d^5\ 4s^1$
Ni: $3d^8\ 4s^2$

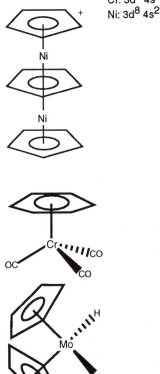

Intrinsic reaction
MX → decomposition products, elements, polymer, or disproportionation.

Extrinsic reaction
MX → decomposition products, oxidation, or hydrolysis.

	Intrinsic reactions		Extrinsic reactions	
	ΔG +ve Thermodynamically stable	ΔG −ve Thermodynamically unstable	ΔG +ve Thermodynamically stable	ΔG −ve Thermodynamically unstable
$\Delta E^{\ddagger}$ small, kinetically labile	Easy to handle and store, as long as its extrinsic reactions do not have ΔG−ve. May be handled in air, otherwise necessary to store in absence of air, e.g. Cr(η-benzene)$_2$ or moisture e.g.SiCl$_4$	Decomposes spontaneously and therefore **cannot be isolated** unless the temperature is lowered or the compound is isolated in an inert gas matrix e.g. XeCl$_2$	Thermodynamically stable even under atmospheric conditions, e.g. NaCl	Kinetically labile towards atmospheric reagents and therefore can only be isolated in the absence of these reagents using vacuum or Schlenk line techniques, e.g. Li$_4$Me$_4$ PMe$_3$
$\Delta E^{\ddagger}$ large, kinetically inert	Thermodynamically stable and kinetically inert, may be handled in air, e.g. SF$_6$.	Thermodynamically unstable but kinetically inert, may be handled in air, e.g. KClO$_4$	Thermodynamically stable and kinetically inert towards atmospheric reagents, may be handled in air, e.g. [Co(NH$_3$)$_6$]$^{3+}$	Thermodynamically unstable but kinetically inert towards atmospheric gases and H$_2$O. Can be stored without precaution, e.g. SiMe$_4$.

Besides reactions with the atmosphere and moisture compounds may also be light or shock sensitive and it may be necessary to take additional precautions to prevent their decomposition.

Steric effects

Steric effects can play an important role in many areas of inorganic chemistry.

1. Unusual co-ordination numbers
Ligands which have large organic pendant groups screen the metal from the approach of other ligands and also limit the number of ligands which can be placed around the metal. Therefore lower co-ordination numbers are favoured.

2. Multiple bond formation
The silicon analogue of ethene Si$_2$H$_4$ cannot be isolated but when the Hs are replaced by bulky R groups, the compounds Si$_2$R$_4$ are formed.

3. Stable radicals
Group IV compounds in their (+3) oxidation state may be isolated because the following dimerization process is hindered if R is large:

$$2SnR_3 \rightleftharpoons R_3Sn-SnR_3$$

4. Alternative co-ordination geometries
The tetrahedron provides a more effective way of arranging four ligands around a metal to minimize the steric repulsions compared with the alternative square planar geometry. Therefore, larger ligands favour the tetrahedral geometry at the expense of the square-planar.

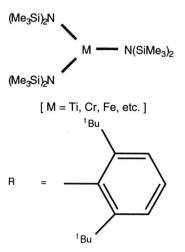

Valence bond

This is a quantum mechanical expression of the Lewis concept of the bond, whereby it is assumed that structures with all the electrons paired in all possible ways dominate the structure wavefunction of the molecule. The final wavefunction is a superposition of all possible perfectly paired canonical structures with those of lowest energy making the main contribution. This superposition is described as *resonance*.

Examples:

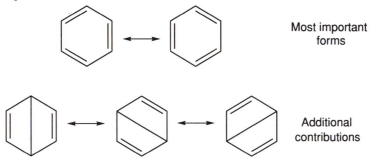

Most important forms

Additional contributions

The technique becomes problematical when there are many structures with the same energy and, for example, a sandwich structure such as ferrocene $Fe(\eta-C_5H_5)_2$ which requires hundreds of resonance structures. Also, if the ground state does not have all the electrons spin paired then simple resonance forms do not give a reliable estimate of the energy.

Example:

Looks like effective resonance stabilization, but ground state is a triplet and the molecule is anti-aromatic

Similarly, O_2 has a triplet ground state.

Valence shell electron pair repulsion (VSEPR)

This is a semi-classical theory which accounts qualitatively for the shapes of most main group molecules. Electron pairs are an important feature of the Lewis description of chemical bonds and VSEPR gives a geometric dimension to the theory by proposing that the most favourable molecular shape is that which minimizes the repulsion between sigma-bonding electron pairs around specific atoms.

The polyhedra which minimize these electron repulsions in a molecule AB_n have shapes which are described in the table below and are illustrated along the diagonals of the table shown on page 90.

AB_n	7	6	5	4	3	2
	Pentagonal bipyramid	Octahedron	Trigonal bipyramid	Tetrahedron	Trigonal plane	Linear

Number of sigma electron pairs	*n* = 7	*n* = 6	*n* = 5	*n* = 4	*n* = 3	*n* = 2

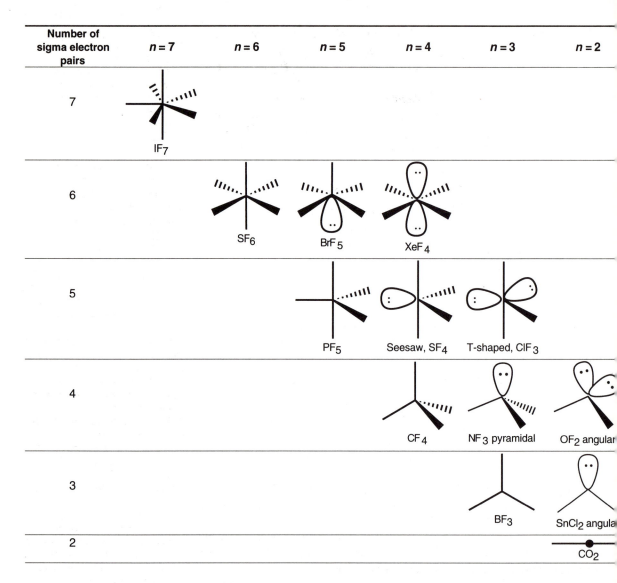

If the atoms B are replaced by lone pairs the polyhedra are retained and an electron pair occupies the vertex which was previously occupied by B. For example, the following molecules SF_6, BrF_5, and XeF_4 all have six σ-electron pairs and have geometries based on the octahedron.

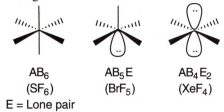

| AB_6 | AB_5E | AB_4E_2 |
| (SF_6) | (BrF_5) | (XeF_4) |

E = Lone pair

The table above shows more examples of this important structural principle for AB_n (*n* = 5 to 2).

More detailed aspects of the shapes are derived by recognizing that bonding pairs and lone pairs do not generate the same extent of repulsion. Specifically the following order of repulsions accounts for the observed geometries:

lone pair–lone pair > lone pair–bond pair > bond pair–bond pair

The molecule adopts its shape, or adjusts its shape, to minimize these repulsions. For example, XeF_4 adopts a *trans*-arrangement of lone pairs rather than a *cis*- arrangement because the lone pair–lone pair repulsions are minimized in the former.

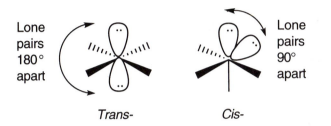

| Lone pairs 180° apart | | Lone pairs 90° apart |

Trans-   *Cis-*

The manner in which the molecular shape is adjusted to minimize repulsion may be illustrated by the following series which have geometries based on the tetrahedron.

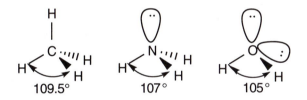

109.5°   107°   105°

In ammonia and water the H–N–H and H–O–H bond angles decrease to reduce the lone pair–bond pair repulsions. The energy gain compensates for the increased bond pair–bond pair repulsions.

Finally, multiple bonds exert a stronger repulsion than single bonds and the molecule adjusts its geometry in an analogous manner. The following molecules illustrate specific applications of this principle:

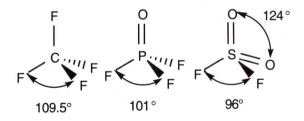

109.5°   101°   96°

Exceptions

1. Alternative geometries are at times observed for AB_n with 5, 7, and 8 co-ordination because they have similar energies to those predicted by VSEPR. Examples, $[SbPh_5]$ and $[InCl_5]^{2-}$ have square-pyramidal not trigonal bipyramidal geometries; $[NbF_7]^{2-}$ (capped trigonal prismatic) and $[NbOF_6]^{3-}$ (capped octahedral) rather than pentagonal bipyramidal.

2. The lone pair in AB_nE is not stereochemically active in the following AB_6E compounds $[TeCl_6]^{2-}$ and $[SbBr_6]^{3-}$ ions which retain a regular octahedral geometry. Similarly although PbO has an infinite layer lattice based on AB_4E square pyramids with E in the axial position PbS has a sodium chloride structure with regular octahedra.

3. Steric effects resulting from bulky ligands can overwhelm lone pair effects. For example, both $SnCl_2$ and $Sn(C_5H_5)_2$ have angular geometries but $Sn(C_5Ph_5)_2$ has a linear geometry.

4. Strong multiple bonding effects can lead to delocalization of the lone pair and consequently it is no longer stereochemically active. Examples, $C(NO_2)_3^-$ is trigonal planar not pyramidal, $[O\{RuCl_5\}_2]^{4-}$ is linear about oxygen because of $p\pi$–$d\pi$ bonding.

5. Co-ordination compounds with incomplete d shells frequently have geometries which do not conform to VSEPR and it is necessary to use alternative ligand field theory arguments to rationalize their structures.

6. The alkaline earth halides MX_2 in the gas phase have linear geometries for M = Be and Ca, but angular geometries for the heavier metals and this has been ascribed to s–d hybridization effects because the d shells are of lower energy than $(n+1)p$.

...ngford, *Inorganic chemistry*, 2nd

...ced inorganic chemistry*, 5th edn.

A. Earnshaw and N.N. Greenwood, *Chemistry of the elements*. Pergamon Press. (1984).

Jack Barrett, *Understanding inorganic chemistry*. Ellis Horwood (1991).

William L. Jolly, *Modern inorganic chemistry*, 2nd edn. McGraw-Hill (1991).

Thermodynamic aspects

W.E. Dasent, *Inorganic energetics*. Cambridge University Press (1982).

D.A. Johnson, *Some thermodynamic aspects of inorganic chemistry*. Cambridge University Press (1982).

Bonding

M.J. Winter, *Chemical bonding*, Oxford Chemistry Primer No. 15. Oxford (1994).

P.A. Cox, *The electronic structure and chemistry of solids*. Oxford University Press (1987).

R. DeKock and H.B. Gray, *Chemical structure and bonding*, Benjamin/Cummings, Menlo Park (1980).

P.W. Atkins, *Quanta*. Oxford University Press (1991).

Periodic table

R.J. Puddephatt and P.K. Monaghan, *The periodic table of the elements*. Oxford University Press (1986).